ABC Transporters in Human Disease

ABC Transporters in Human Disease

Karobi Moitra

National Cancer Institute (at Frederick)

COLLOQUIUM SERIES ON THE GENETIC BASIS OF HUMAN DISEASE #1

MORGAN & CLAYPOOL LIFE SCIENCES

ABSTRACT

The ATP-binding cassette (ABC) transporter genes are ubiquitous in the genomes of all vertebrates so far studied. The human ABC transporter superfamily contains 48 genes, subdivided into 7 sub-families ranging from A to G (based on sequence homology of their nucleotide binding domains). The ABC proteins encoded by these genes are ATP-driven transmembrane pumps, some of which possess the capacity to efflux harmful toxic substances and therefore play a key role in xenobiotic defense. ABC proteins have been evolutionarily conserved from bacteria to humans and multiple gene duplication and deletion events in the ABC genes indicate that the process of gene evolution is still ongoing. Polymorphisms and variations in these genes are linked to variations in expression, function, drug disposition, and drug response. Single nucleotide polymorphisms (SNPs) in these genes could be markers of individual risk for adverse drug reactions or susceptibility to complex diseases. The pharmacogenetics of this unique family of transporters is still under study; however, in the context of human health, it is a well-known fact that variations in these transporters are the un-derlying cause for several human diseases including cystic fibrosis, *Pseudoxanthoma elasticum* (PXE), and X-linked adenoleukodystrophy (X-ALD).

KEYWORDS

ABC transporters, human disease, ATP, evolution, cystic fibrosis, PXE, X-ALD

Contents

CHAPTER 1

Introduction to the Human ATP-Binding Cassette (ABC) Transporter Superfamily

1.1   ABC GENE AND PROTEIN ORGANIZATION

ATP-binding cassette transporters (ABC-transporters) catalyze the vectorial transport of a great variety of substrates across biological membranes. They are intricate molecular systems that have representatives in all extant phyla from prokaryotes to humans. They are one of the largest and ancient protein superfamilies [1] and are able to power the translocation of substrates (including sugars, amino acids, metal ions, peptides/proteins, and a large number of hydrophobic compounds) across biological membranes, often against a concentration gradient, by the hydrolysis of ATP [2].

1.2   DOMAIN STRUCTURE AND ORGANIZATION OF ABC TRANSPORTER GENES AND PROTEINS

The human genome encodes 48 ABC transporter genes, several of which are medically relevant. The 48 ABC proteins in humans are divided into seven subfamilies, from A to G, based on structural arrangements and phylogenetic analysis [3, 4]. Proteins are classified as ABC transporters on the basis of their sequence and organization of ATP-binding domain(s), also known as nucleotide-binding domains (NBDs) or nucleotide-binding-folds (NBFs). The NBFs contain characteristic motifs (the Walker A motif and Walker B motif) found in all ATP-binding proteins. The ABC genes also contain an additional element, the signature motif (C-loop) that is located upstream of the Walker B site [5], shown in Figure 1. The functional ABC protein typically contains two NBFs and two transmembrane domains (TMDs). The TM domains consist of 6–12 membrane-spanning alpha-helices (Figure 1 and Figure 2) which are mainly responsible for determining substrate specificity. Mutations in the ABC genes can cause or contribute to several human genetic disorders (Table 1) including neurological disease, retinal degeneration, cholesterol/bile transport defects, cystic fibrosis, anemia and differential drug response [6].

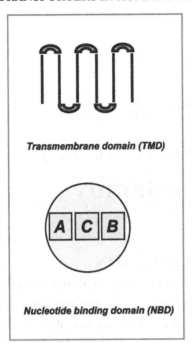
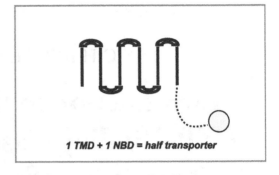
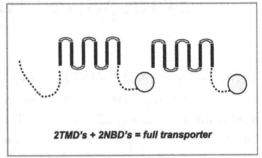

FIGURE 1: Domain organization in ABC transporters. ABC transporters consist of transmembrane domains (TMDs) typically containing six transmembrane α helices and nucleotide binding domains (NBDs, yellow). NBD's contain Walker A, Walker B and C-loop functional sites (green squares). ABC full transporters contain two NBDs and two transmembrane domains. Half-transporters typically contain one TMD and one NBD.

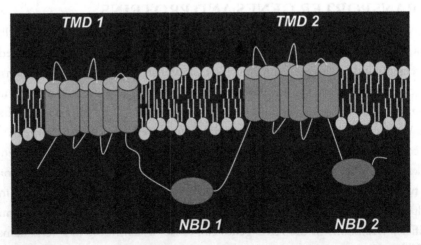

FIGURE 2: Schematic representation of a typical ABC full transporter. The two TMD's (TMD1 and TMD2) consist of 6 transmembrane helices (orange). Two intracellular NBD's are responsible for ATP hydrolysis (red ovals).

TABLE 1: Diseases and phenotypes caused by ABC genes.

GENE	MENDELIAN DISORDER	COMPLEX DISEASE
ABCA1	Tangier disease, FHDLD [a]	
ABCA4	Stargardt/FFM, RP, CRD, CD	AMD
ABCB1	Ivermectin susceptibility	Digoxin uptake
ABCB2	Immune deficiency	
ABCB3	Immune deficiency	
ABCB4	PFIC3	ICP
ABCB7	XLSA/A	
ABCB11	PFIC2	
ABCC2	Dubin-Johnson Syndrome	
ABCC6	Pseudoxanthoma elasticum	
ABCC7	Cystic Fibrosis, CBAVD	Pancreatitis, bronchiectasis
ABCC8	FPHHI	
ABCD1	ALD	
ABCG5	Sitosterolemia	
ABCG8	Sitosterolemia	

[a] FHDLD, familial hypoapoproteinemia; FFM, fundus flavimaculatis; RP, retinitis pigmentosum 19; CRD, cone-rod dystrophy; AMD, age-related macular degeneration; PFIC, progressive familial intrahepatic cholestasis; ICP, intrahepatic cholestasis of pregnancy; XLSA/A, X-linked sideroblastosis and anemia; CBAVD, congential bilateral absence of the vas deferens; FPHHI, Familial persistent hyperinsulinemic hypoglycemia of infancy; ALD, adrenoleukodystrophy.

1.3   MECHANISM OF TRANSPORT

The intricate mechanisms by which ABC transporters move solutes across membranes are not clear. For all intents and purposes, it seems that the initiation of the transport cycle starts when the substrate becomes bound to the transmembrane domains of the transporter. This induces a conformational change in the TMDs which is then transmitted to the NBDs. The NBDs can then initiate ATP binding and ATP hydrolysis. ATP hydrolysis powers the transport of substrates across the plasma membrane [7, 8, 9] (Figures 3a and 3b).

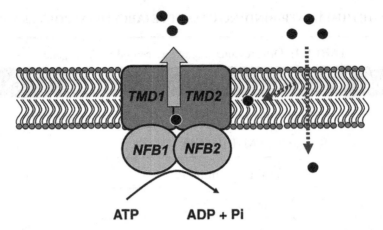

FIGURE 3A: ABC transporters are driven by ATP. ABC transporters efflux substrates with the power provided by ATP hydrolysis. ABC, ATP-binding cassette; NBF, nucleotide-binding fold; TMD, trans-membrane domain. Reproduced with permission from Nature Publishing Group (NPG). Moitra K et al. *Clinical Pharmacology and Therapeutics* 2011, 89(4): 491–502 [8].

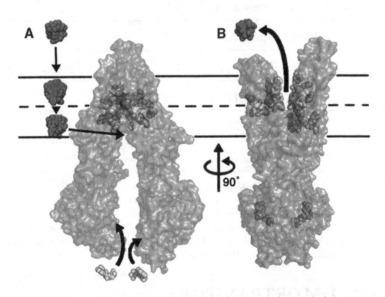

FIGURE 3B: Model of substrate transport by P-gp. (A) Substrate (magenta) partitions into the bilayer from outside of the cell to the inner leaflet and enters the internal drug-binding pocket through an open portal. The residues in the drug-binding pocket (cyan spheres) interact with QZ59 compounds and verapamil in the inward-facing conformation. (B) ATP (yellow) binds to the NBDs causing a large conformational change presenting the substrate and drug-binding site(s) to the outer leaflet and/or extracellular space. In this model of P-gp, which is based on the outward-facing conformation of MsbA and Sav1866, exit of the substrate to the inner leaflet is sterically occluded, which provides unidirectional transport to the outside. Reproduced with permission from AAAS, S G Aller et al. *Science* 2009;323:1718–1722 [9].

1.4   CLASSIFICATION OF ABC TRANSPORTER SUBFAMILIES IN HUMANS

All human and mouse ABC genes have standard nomenclature, developed by the Human Genome Organization (HUGO). A list of the 48 human ABC genes [10] is shown in Table 2. Analysis of the complete human genome sequence has failed to reveal any other additional genes [11, 12]. Using sequence alignment strategies mainly by aligning the amino acid sequences of the NBDs and performing phylogenetic analysis, the existing eukaryotic genes can be grouped into seven major subfamilies. There is an eighth subgroup, ABCH, which contains members in insects, *D. discoideum* (slime mold), and fish. A number of genes in yeast and protozoa also do not fit into these subfamilies. Some of the subfamilies can be additionally divided into subgroups. For example, the ABCB subfamily contains both full transporters (with 2 NBDs and 2 TMDs) and half transporters (with 1 NBD and 1 TMD). The situation in plants is that they possess fused, full transporter forms of ABCD family genes. The organization of the TM domains and/or phylogenetic analysis can be used to divide the subfamilies [13]. Figure 4 depicts the phylogenetic tree of human ABC transporters.

1.4.1   ABCA

The human ABCA subfamily comprises 12 full transporters that can be further divided into two subgroups based on phylogenetic analysis and intron structure [14, 15]. The first group includes seven genes dispersed on six different chromosomes (*ABCA1*, *ABCA2*, *ABCA3*, *ABCA4*, *ABCA7*, *ABCA12*, and *ABCA13*), whereas the second group contains five genes (*ABCA5*, *ABCA6*, *ABCA8*, *ABCA9*, and *ABCA10*) arranged in a cluster on chromosome 17q24.3. The ABCA1 protein is involved in disorders of cholesterol transport and HDL biosynthesis. The ABCA4 protein is thought to transport vitamin A derivatives in the outer segments of rod photoreceptor cells and therefore performs a crucial step in the vision cycle. Evolutionary studies of ABCA genes seem to indicate that these genes arose as half transporters that subsequently duplicated, and certain sets of ABCA genes were lost in different eukaryotic lineages [16].

1.4.2   ABCB (MDR/TAP)

The ABCB subfamily consists of both full transporters and half transporters. Four full transporters and seven half transporters have been described in this subfamily. *ABCB1* (*MDR/PGY1*) was the first human ABC transporter to be cloned and characterized by its ability to confer a multi-drug resistant phenotype to cancer cells. Other important ABCB proteins include ABCB4 and ABCB11 proteins that are both located in the liver and involved in the secretion of bile acids. The *ABCB2* (*TAP1*) and *ABCB3* (*TAP2*) genes are half transporters that form a heterodimer to transport peptides into the ER that are presented as antigens by the class I HLA molecules. The closest homolog

TABLE 2: List of human ABC genes, subfamilies, chromosomal locations and functions

SYMBOL	SUBFAMILY	ALIAS	LOCATION	FUNCTION	ANTICANCER DRUG TRANSPORT
ABCA1	A	ABC1	09q31.1	Cholesterol efflux onto HDL	
ABCA2	A	ABC2	09q34.3	Drug transport	Yes
ABCA3	A	ABC3	16p13.3	Surfactant secretion? Drug resistance	Yes
ABCA4	A	ABCR	01p21.3	N-Retinylidiene-PE efflux	
ABCA5	A	ABC13	17q24.3		
ABCA6	A		17q24.3		
ABCA7	A	ABCX	19p13.3		
ABCA8	A		17q24.3		
ABCA9	A		17q24.3		
ABCA10	A		17q24.3		
ABCA12	A		02q34		
ABCA13	A		07p12.3		
ABCB1	B	MDR1	07q21.12	Multidrug resistance	Yes
ABCB2	B	TAP1	06p21	Peptide transport	
ABCB3	B	TAP2	06p21	Peptide transport	
ABCB4	B	PGY3	07q21.12	Phosphotidyl Choline & drug transport	Yes

ABCB5	B	ABC19	07p21.1	Drug transport	Yes
ABCB6	B	ABC14	02q35	Iron transport	
ABCB7	B	ABC7	Xq21-22	Fe/S cluster transport	
ABCB8	B	ABC22	07q36.1		
ABCB9	B	ABC23	12q24.31		
ABCB10	B		01q42.13		
ABCB11	B	SPGP	02q24.3	Bile salt transport, drug transport	Yes
ABCC1	C	MRP1	16p13.12	Drug resistance	Yes
ABCC2	C	MRP2	10q24.2	Organic anion efflux, drug transport	Yes
ABCC3	C	MRP3	17q21.33	Drug resistance	Yes
ABCC4	C	MRP4	13q32.1	Nucleoside transport, drug transport	Yes
ABCC5	C	MRP5	03q27.1	Nucleoside transport, drug transport	Yes
ABCC6	C	MRP6	16p13.12		Yes
ABCC7	C	CFTR	07q31.31	Chloride ion channel	
ABCC8	C	SUR	11p15.1	Sulfonylurea receptor	
ABCC9	C	SUR2	12p12.1	K(ATP) channel regulation	
ABCC10	C	MRP7	06p21.1	Drug transport	Yes
ABCC11	C	MRP8	16q12.1	Drug transport	Yes

TABLE 2: (*continued*)

SYMBOL	SUBFAMILY	ALIAS	LOCATION	FUNCTION	ANTICANCER DRUG TRANSPORT
ABCC12	C	MRP9	16q12.1		
ABCD1	D	ALD	Xq28	VLCFA transport regulation	
ABCD2	D	ALDL1	12q11		
ABCD3	D	PXMP1	01p22.1		
ABCD4	D	PMP69	14q24.3		
ABCE1	E	OABP	04q31.31	Oligoadenylate binding protein	
ABCF1	F	ABC50	06p21.1		
ABCF2	F	ABC28	07q36.1		
ABCF3	F	ABC25	03q27.1		
ABCG1	G	White	21q22.3	Cholesterol transport?	
ABCG2	G	ABCP	04q22	Toxin efflux, drug resistance	Yes
ABCG4	G	White2	11q23		
ABCG5	G	White3	02p21	Sterol transport	
ABCG8	G	White4	02p21	Sterol transport	

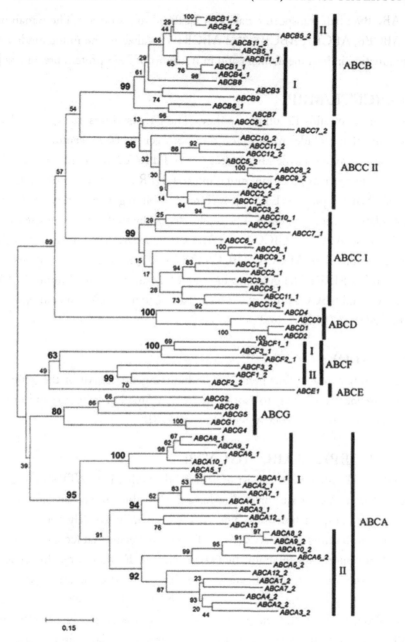

FIGURE 4: Phylogenetic tree of the human ABC genes. Reproduced with permission from Elsevier Sciences Ltd, ABC Proteins from Bacteria to Man, 2003, Eds Holland, Cole, Kuchler and Higgins, ISBN 0-12-352551-9 [10] and also with permission from Michael Dean.

of the TAPs, ABCB9 a half transporter, has been localized to lysosomes. The remaining four half transporters, ABCB6, ABCB7, ABCB8, and ABCB10, localize to the mitochondria, where they are thought to function in iron metabolism and in transport of Fe/S protein precursors [13].

1.4.3 ABCC (CFTR/MRP)

The ABCC subfamily contains 12 full transporters. These transporters, as expected, have a diverse functional spectrum that includes roles in ion transport, cell-surface receptor, and toxin secretion activities. The cystic fibrosis transmembrane receptor (CFTR/ABCC7) protein is a chloride channel that plays a role in exocrine secretion; mutations in CFTR cause the disease cystic fibrosis [17]. The ABCC8 and ABCC9 proteins bind sulfonylurea and can regulate potassium channels that are involved in modulating insulin secretion. The remainder of the subfamily comprises of nine MRP-related genes. Of these, ABCC1, ABCC2, and ABCC3 transport drug conjugates to glutathionine and other organic anions. The ABCC4, ABCC5, ABCC11, and ABCC12 proteins happen to be smaller than the other MRP1-like gene products and lack an N-terminal domain (TMD0) [18]. Finally, the ABCC4 and ABCC5 proteins confer resistance to nucleosides including 9-(2-phospho-nylmethoxyethyl)adenine (PMEA) and purine analogs.

1.4.4 ABCD (ALD)

The ABCD subfamily contains four genes in the human genome. All of the genes encode half transporters that are located in the peroxisome, where they function as homo- and/or heterodimers in the regulation of very long chain fatty acid transport [13].

1.4.5 ABCE (OABP) and ABCF (GCN20)

The ABCE and ABCF subfamilies consist of gene products that have ATP-binding domains and are clearly derived from ABC transporters but have no TM domain and are not known to be involved in membrane transport functions. The ABCE subfamily is solely composed of a protein called oligo-adenylate-binding protein (OABP), which is a molecule that recognizes oligo-adenylate and is produced in response to infection by certain viruses. Furthermore, this gene is found in multicellular eukaryotes but not in yeast, suggesting that it is part of innate immunity. In regards to the ABCF gene family, each gene contains a pair of NBDs. The best-characterized member, the *S. cerevisiae GCN20* gene product, mediates the activation of the eIF-2a kinase [19], and a human homolog, ABCF1, is associated with the ribosome and appears to play a similar role [20].

1.4.6 ABCG (White)

The human ABCG subfamily is made up of six "reverse" half transporters that have an NBD at the N terminus and a TM domain at the C terminus. The most intensively studied ABCG gene is the *white* locus of *Drosophila*. The white protein (along with brown and scarlet) transports precursors of eye pigments (guanine and tryptophan) in the eye cells of the fly [21]. The mammalian ABCG1 protein is involved in cholesterol transport regulation [22]. Other ABCG genes include *ABCG2*, which is a drug-resistance gene; *ABCG5* and *ABCG8* that code for transporters of sterols in the intestine and liver; *ABCG3*, to date exclusively found in rodents; and the *ABCG4* gene is expressed predominantly in the liver. The functions of the last two genes are currently unknown.

. . . .

CHAPTER 2

Evolution of ABC Transporters

2.1 EVOLUTION OF ABC TRANSPORTERS BY GENE DUPLICATION

ABC transporter genes are intricately involved in human genetic diseases and contribute to important drug transport phenotypes. Vertebrate evolution has largely been driven by the duplication of genes that allow for the acquisition of new functions. Multiple gene duplication and deletion events have been identified in the ABC genes in diverse lineages, indicating that the process of gene evolution is still ongoing [23]. Studies that involve a number of gene families have revealed three possible scenarios where two identical genes can be generated by duplication [24, 25]. The most common scenario is that one of the copies is silenced by mutations or completely deleted (*complete deletion*). In another scenario, one copy of the gene retains its original function, while the other acquires a new function. This is termed as *neo-functionalization*. A third possibility is *sub-functionalization*, which may occur when the original function of the gene is separated (either spatially or by timing of expression) between the duplicates. The ABC transporter genes are highly conserved in nature and thus can be assumed to encode proteins whose functions have changed very little over time. Moreover, lineage-specific genes are more likely to encode specialized transporters. Over 94% of all human ABC genes have an ortholog in each studied mammal, and 85% of the genes are orthologous in chicken, 77% in zebrafish, and 40% in *Ciona intestinalis*. Studies on the functions of ABC genes in multiple species are likely to shed insight into the process of gene evolution [23]. Gene duplication is an important process for supplying raw genetic material in biological evolution [26]. Genomic sequence data provide substantial evidence for the abundance of duplicated genes in all surveyed organisms [25]. Duplication occurs in an individual- it can be fixed or lost in the population, similar to a point mutation. If a new allele consisting of duplicate genes is selectively neutral, it only has a very small probability of being fixed in a diploid population [27] this would suggest that many duplicated genes will be lost. When genes become fixed, the long-term evolutionary fate of duplication will be determined by the function of the duplicate genes. Genes involved in physiological processes that vary greatly among species (e.g., immunity, reproduction, and sensory systems) probably have high rates of *gene birth* (emergence of a new gene) and *gene death* (loss of a functional gene). Birth and death of genes is a common motif in gene family and genome evolution [28, 29] which is highly relevant to the evolutionary process occurring in the ABC transporter gene family.

We have previously shown that there are a few locations in the mammalian genome where ABC genes form clusters, and the process of birth and death is more intense in these regions than elsewhere [30]. A cluster of closely related genes can in fact give rise to new genes by unequal crossing over, as found in the *ABCB5, ABCB6, ABCB8, ABCB9, ABCB10* cluster.

2.2   CONSERVATION OF ABC GENES IN VERTEBRATES

Most of the ABC genes are conserved across placental mammals with a few notable exceptions. In the ABCA subfamily, *ABCA1* is highly conserved. In humans, this transporter plays a pivotal role in the cholesterol transport from peripheral cells into high-density lipoprotein particles [31]. *ABCA3* is also highly conserved and has given rise to a number of *ABCA3*-like genes. *ABCA4* is expressed in mammalian photoreceptor cells where it possibly facilitates the outward transport of retinoid-lipid complexes [32]. Most of the genes in the B subfamily (*ABCB1, ABCB2, ABCB3, ABCB7, ABCB8, ABCB9, ABCB10,* and *ABCB11*) are also conserved across the placental mammals. The ABCC family consists of transporters with varied functions. *ABCC1, ABCC3, ABCC4–ABCC7, ABCC9,* and *ABCC10* are ubiquitously present in placental mammals. The ABCD subfamily codes for half transporter genes expressed in the peroxisomes. Some of these D family transporters are needed for the metabolism of very long chain fatty acids [33]. The ABCE and F family genes are extraordinary because they only have 2 ATP-binding domains but no TM domains. These genes are also highly conserved across the placental mammals. The G family of half transporters is unique in that they have an NBD at the N terminus of the protein and one TMD at the C terminus. All of them except *ABCG3* are present in all the vertebrates studied. *ABCG3* is closely related to *ABCG2* but is found only in the rodents where it could have evolved to perform some kind of specialized function.

2.3   BIRTH AND DEATH EVOLUTION IN ABC TRANSPORTERS

Birth-and-death evolution is a form of independent evolution in which new genes are created by repeated gene duplication. Some duplicate genes may continue in the genome for a long time, while others may be deleted or become non-functional [34, 35]. A number of ABC genes have arisen by the process of duplication, leading to gene birth. An active gene duplication process occurring in the ABC transporter family in vertebrates has been documented and includes apparently ancient events such as the whole-genome duplication in fish and also more recent events such as *ABCB1* duplications in rodents and opossum and the duplication of the *ABCG3* gene which is specific to rodents [23].

2.3.1   Gene Birth

The process of gene birth is exemplified by the *ABCG* transporter genes shown in Figure 5. ABCG2 (in humans) is a half-transporter which is thought to function as a homodimer. The physiologi-

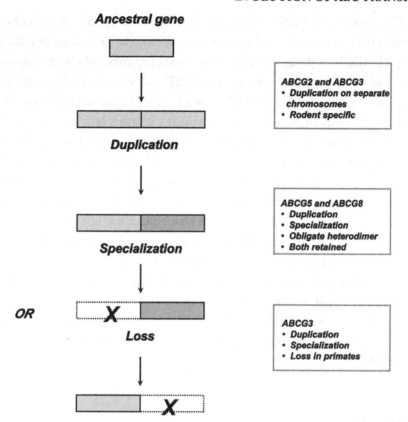

Ancestral gene

Duplication

ABCG2 and ABCG3
• *Duplication on separate chromosomes*
• *Rodent specific*

Specialization

ABCG5 and ABCG8
• *Duplication*
• *Specialization*
• *Obligate heterodimer*
• *Both retained*

OR   **X**

Loss

ABCG3
• *Duplication*
• *Specialization*
• *Loss in primates*

X

FIGURE 5: The evolution of G subfamily transporters by the process of gene birth and gene death. The 'G' subfamily of transporters initially evolved from an ancestral transporter gene (light pink). In rodents ABCG2 and ABCG3 duplicated and became localized on separate chromosomes. ABCG5 underwent duplication to give rise to ABCG8 (green). In primates ABCG3 was lost (gene death).

cal substrate of ABCG2 is unknown but it can function as a multi-drug transporter conferring resistance to anthracycline anticancer drugs such as mitoxantrone [36–39]. The *ABCG3* gene is related to *ABCG2*. This gene is primarily expressed in murine hematopoietic cells and contains no ortholog in the human genome, although rodents appear to have an orthologous sequence. The *ABCG3* gene has an unusual nucleotide binding domain that has alternative residues in several conserved positions, which may suggest that it might either fail to bind or hydrolyze ATP [40]. From the neighbor-joining tree, we can see that *ABCG3* genes are present only in rodents and are very similar to the *ABCG2* rodent genes. Presumably, *ABCG3* arose after the rodents split but the phylogenetic distance indicates that it may have evolved more rapidly to acquire a unique functional role in rodents. Other proposed duplication events leading to gene birth include duplications of

ABCB1 and *ABCB4*, along with *ABCG5* and *ABCG8* (Figure 5). ABCB1 (P-glycoprotein) is a full transporter and is the best characterized multidrug pump, and the gene maps to 7q21.1 in the human genome [41]. ABCB1 is thought to play an important physiological role of removing toxic metabolites from cells. *ABCB4* is adjacent to *ABCB1* at 7q21.1. It encodes a full transporter which has high sequence homology to ABCB1. ABCB4 is expressed mainly in the bile canicular membrane of the liver and in vivo experiments have shown that it can transport phosphotidyl choline from the inner to the outer leaflet of membranes [42, 43]. Thus, we can see that both these genes arose from an 'ancestral gene', underwent 'duplication' but then became 'specialized' to perform divergent functions (Figure 5). ABCG5 and ABCG8 are half transporters which transport sterols. The *ABCG5* gene maps in humans to chromosome 2p21 and is adjacent to the *ABCG8* gene [44, 45]. These two transporters seem to form a functional heterodimer and appear to be regulated by the same promoter [46]. Both *ABCG5* and *ABCG8* genes are mutated in families that have 'sitosterolemia,' a disease characterized by the defective transport of plant (and fish) sterols and cholesterol [47]. Therefore, *ABCG5* and *ABCG8* also arose from an 'ancestral gene', underwent 'duplication' but unlike ABCB1 and ABCB4 did not become specialized to perform divergent functions but became 'functionally obligate heterodimers' (Figure 5) performing a common function (i.e., sterol transport). In the above examples, both the genes in the pair were retained, thus giving rise to new genes or 'gene birth.'

2.3.2   Gene Death

The highest rate of gene duplication followed by 'gene death' can be seen in the ABCA subfamily of ABC transporters. In the zebrafish genome, only 7 of the *ABCA* genes are observed [48]. The most notable expansion of *ABCA* genes in mammals are the *ABCA3*-like genes—*ABCA14*, *ABCA15*, *ABCA16*. These genes form a cluster on mouse chromosome 7. *ABCA17* is present in rodent genomes where it is a sperm-specific transporter regulating intracellular lipid metabolism [49]. These genes are highly expressed in mice testes suggesting a role for these genes in testicular function and/or sperm maturation. From the tree (Figure 6), it can be seen that some of these genes are present in other mammals such as rabbit, horse, and elephant, however, not in the primates. All these genes are believed to have arisen by duplication of *ABCA3* and became 'specialized' in the non-primate mammals but ultimately were lost or have undergone 'gene death' (Figure 6) in the primate lineages. Additionally, a number of 'pseudogenes' (non-coding DNA sequences that resemble a functional gene) are also present in the ABC transporter gene family. Pseudogenes of *ABCA3*, *ABCC6*, and *ABCD1* can be found that have resulted from either full or partial duplication. One duplicated copy of the gene becomes degenerated due to insertions and deletions of different sizes [48] providing ample evidence for gene duplication followed by 'gene death' in this family of transporters.

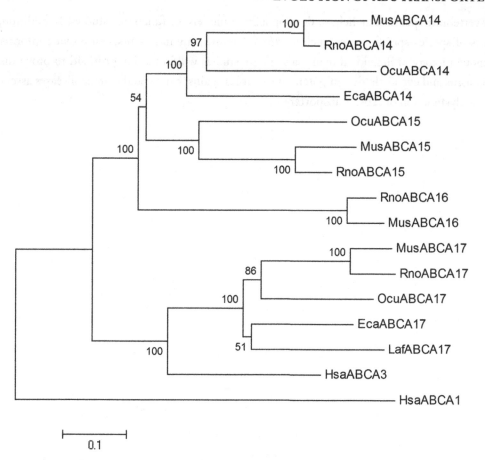

FIGURE 6: Phylogenetic (neighbor-joining) tree of ABCA transporters. The tree depicts expansion followed by 'death' of the ABCA3-like genes in selective mammalian lineages.

2.4   CONCLUSIONS

The ABC transporter genes seem to be highly conserved in mammals. Birth and death of genes can keep the number of genes in a gene family relatively constant or in a state of balance. This has been shown for the major histocompatibility complex [50] and may be applicable to the ABC genes. Duplication has played a pivotal role in the evolution of new gene functions. Duplicated genes may have several evolutionary fates corresponding to either the process of gene birth or gene death. ABC genes are involved in various human diseases and studying the evolution of these genes in other vertebrate species may help to develop animal models for functional studies. In addition, the analysis of species-specific genes, such as *ABCG3* found only in rodents, can provide information on unique biology in individual mammals. These studies will in the long run aid to probe disease mechanisms and even help design potential therapies against cancer and genetic diseases associated with the dysfunction of ABC transporters.

.   .   .   .

CHAPTER 3

Overview of ABC Transporters in Human Disease

3.1   ABC GENES ASSOCIATED WITH HUMAN DISEASE

At least 14 ABC genes have been linked to specific human genetic disorders [51] (Table 1). Many of these were located by a positional cloning approach (a method of gene identification by the isolation of partially overlapping DNA segments from genomic libraries to progress along the chromosome toward a specific gene).

3.2   ABCA1 AND TANGIER DISEASE

The causative gene in Tangier disease is ABCA1. Tangier disease is named for the island of Tangier in the Chesapeake Bay where the first affected patients were discovered. It is a disorder of cholesterol transport between tissues and the liver, mediated by binding of the cholesterol onto high-density lipoprotein (HDL) particles [52–56]. It is characterized by low levels of HDL ('good cholesterol') in the blood, neuropathy, and enlarged orange-colored tonsils. The gene locus was mapped to chromosome 9—the ABCA1 gene.

3.3   THE ABCA4 GENE AND EYE DISORDERS

The ABCA4 (ABCR) gene has been found to map to chromosome 1p21.3 and is expressed exclusively in photoreceptors, where it is believed to transport retinol (vitamin A)/phospholipid derivatives from the photoreceptor outer segment disks into the cytoplasm [57–59]. Mutations in the ABCA4 gene have been associated with multiple spectrums of ophthalmic disorders [60]. Complete loss of function of ABCA4 leads to retinitis pigmentosa. Patients with at least one missense allele have Startgardt disease [61–63]. Another disorder is Startgardt disease which is characterized by juvenile to early adult-onset macular dystrophy with loss of central vision [64] . To date, there are over 500 mutations in the gene encoding ABCA4 which are associated with a broad spectrum of related autosomal recessive retinal degenerative diseases. [65]. Age-related macular degeneration (AMD) patients also have an increased frequency of ABCA4 mutation carriers [66]. AMD patients display a wide variety of phenotypes that include the loss of central vision, after 60 years of age. The postulated mechanism of disease is that there may be an abnormal accumulation of retinoids in the eye due to ABCA4 deficiency.

3.4 ABCB1 IN IVERMECTIN SUSCEPTIBILITY AND CANCER DRUG RESISTANCE

The ABCB1 (PGP/MDR1) gene maps to chromosome 7q21.1, it was the first human ABC transporter cloned and characterized through its ability to confer a multidrug resistance phenotype to cancer cells that had developed resistance to chemotherapy drugs [67–70].

The function of MDR1 was first discovered in mice, MDR1a knockouts which were susceptible to the anti-parasitic drug ivermectin. ABCB1 has broad substrate specificity and has been demonstrated to transport of hydrophobic substrates including drugs such as colchicine, etoposide (VP16), adriamycin, and vinblastine as well as lipids, steroids, xenobiotics, and peptides. The protein also plays an important role in removing toxic metabolites from cells and also affects the pharmacology of the drugs that are substrates. Pgp is also highly expressed in hematopoietic stem cells, where it may serve to protect these cells from toxins [71, 72]. Over 3300 SNPs (single nucleotide polymorphisms—a DNA variation in a single base) have been documented for human MDR1. Out of these SNPs, a synonymous SNP found in exon 26, C3435T was in some cases found to associate with altered Pgp activity in conjunction with a haplotype (haplotype—a combination of alleles at adjacent positions which may be transmitted together) with reduced functionality [73]. It was found that this silent mutation could induce subtle changes in conformation of the transporter, probably at sites where drugs and modulators interact. Based on this observation, it was hypothesized that when frequent codons are changed to rare codons in a cluster of frequently used codons the timing of co-translational folding is affected which may result in altered function [73]. This finding could be clinically important taking into consideration that diverse populations have different frequencies of this SNP and also of the reduced function haplotype C1236T–G2677T–C3435T.

3.5 TAP1 AND TAP2 IN IMMUNE SYSTEM DISORDERS

The TAP1 (ABCB2) and TAP2 (ABCB3) genes are on chromosome 6p21.3 and comprise of 10 kb each arranged in a head to head orientation in the HLA (human leukocyte antigen) gene complex. They are half transporters that form a heterodimer which can transport peptides into the ER, where they can be complexed with class I HLA molecules for presentation on the cell surface [74–76]. TAP expression is required for the stable expression of class I proteins [77]. Studies have demonstrated that the TAP complex preferentially transports 9–12 amino acid peptides [78] with a preference for Phe, Leu, Arg, and Tyr at the C terminus, similar to the specificity of the HLA class I proteins [78, 79]. An important feature of this heterodimeric transporter is the non-equivalence of its motor domains. The C-loop in TAP2 is degenerated to LAAGQ in contrast to the canonical signature motif (LSGGQ) found in TAP1. It has been hypothesized that one NBD hydrolyzes ATP to provide energy for the transport process while the other NBD acts as a regulatory unit for the ATPase activity [80]. Patients with inherited immunodeficiency because of TAP1 mutations have been described [81]. Bare lymphocyte syndrome (BLS) is one such disease where the cell surface expression of MHC is drastically reduced [82]. The reduced expression is manifested by granulomatous skin lesions and recurrent bacterial infections. In the BLS 1 subgroup of patients, the disease can be traced back to mutations in TAP1 or TAP2.

3.6   ABCB7 IN SIDEROBLASTIC ANEMIA AND ATAXIA

The ABCB7 gene maps to chromosome Xq21–q22. The protein is localized to the mitochondria [83]. ABCB7 gene is closely related to the ABCB6 gene, both of which are half transporters. The ABCB7 gene is found to be mutated in patients with X-linked sideroblastic anemia and ataxia (XLSA/A) [84, 85]. XLSA/A is a recessive disorder characterized by infantile to early childhood onset of non-progressive cerebellar ataxia and mild anemia with hypochromia (less hemoglobin) and microcytosis (red blood cells are smaller than normal).

3.7   ABCB11 IN PATIENTS WITH PROGRESSIVE FAMILIAL INTRAHEPATIC CHOLESTASIS, TYPE 2 (PFIC2)

The ABCB11 (BSEP/SPGP) gene encodes a full transporter protein located principally in the liver and the gene maps to 2q24.3 [86, 87]. The protein participates in the secretion of bile salts such as taurocholate and localizes to bile canalicular membrane of the liver [85]. Mutations in ABCB11 are found in patients with progressive familial intrahepatic cholestasis, type 2 (PFIC2) [88]. More than 30 mutations have been identified in the BSEP gene in patients with BSEP deficiency syndrome [89].

3.8   ABCC2 AND DUBIN–JOHNSON SYNDROME

The ABCC2 (MRP2/cMOAT) gene maps to chromosome 10q24 and is expressed in canalicular cells in the liver [90]. It was found that the ABCC2 gene is mutated in patients with Dubin–Johnson syndrome, a human disorder of organic ion transport [91, 92]. Dubin–Johnson syndrome is characterized by the increased concentration of bilirubin glucuronisides in the blood (hyperbilirubinemia) and the deposition of dark pigment in the hepatocytes. Sequence variants in ABCC2 in patients with Dubin–Johnson syndrome include splice site mutations, missense mutations, deletion mutations, and nonsense mutations leading to premature stop codons [93]. ABCC2 overexpression can confer drug resistance to cells, but the physiological importance of this observation is not clear [94].

3.9   ABCC6 AND PXE

The ABCC6 (MRP6) gene maps to 16p13.1, it is chiefly expressed in the liver and kidney. It is mutated in *pseudoxanthoma elasticum* (PXE), a recessive genetic disorder. PXE is a hereditary disorder that primarily affects the elastic tissues in the skin, eyes, and blood vessels. The skin abnormalities are caused by mineralization of the connective tissue and lead to loss of elasticity and enhanced calcification of elastic fibers. Calcification in the eye may cause macular degeneration and vision loss, and in blood vessels contribute to cardiovascular disease. Analysis of the ABCC6 gene for variants has identified a number of common polymorphisms including missense alleles [95]. The R1141X mutation is one of the most common ABCC6 mutation in PXE patients of European descent, and this variant has been found at levels approaching 1% in these populations. An association of this variant with premature atherosclerotic vascular disease has been reported [96].

3.10   ABCC7/CFTR AND CYSTIC FIBROSIS

The CFTR (ABCC7) gene maps to chromosome 7q31.2 and is a chloride channel that is expressed in exocrine tissues, such as the sweat duct, pancreas, intestine, and kidney. The gene is mutated in the recessive genetic disease cystic fibrosis [97–99]. Cystic fibrosis (CF) is one of the most common fatal childhood diseases in Caucasian populations and is characterized by defective exocrine activity of the lung, pancreas, sweat ducts, and intestine [100]. Through the CFTR protein bacterial toxins, such as cholera toxin and those from *Escherichia coli*, may cause increased fluid flow in the intestine resulting in diarrhea. Based on this observation, several researchers have proposed that the CF mutations have been selected for in response to this disease(s). This hypothesis is supported by studies showing that CF homozygotes fail to secrete chloride ions in response to a variety of stimulants; and that mice in heterozygous null animals show reduced intestinal fluid secretion in response to cholera toxin [101].

3.11   ABCC8 (SUR1) AND FAMILIAL PERSISTENT HYPERINSULINEMIC HYPOGLYCEMIA

The ABCC8 (SUR1) gene maps to chromosome 11p15.1 and codes for a full transporter. The gene encodes a high-affinity receptor of the drug sulfonylurea. Sulfonylureas are a class of drugs commonly used to increase insulin secretion in patients with non-insulin-dependent diabetes. These drugs bind to the ABCC8 protein to inhibit a potassium channel K (ATP) associated with the ABCC8 protein. Familial persistent hyperinsulinemic hypoglycemia of infancy (PHHI) was found to be an autosomal recessive disorder. Patients with this disease show unregulated insulin secretion and various mutations in the SUR1 gene are found in PHHI families [102]. The ABCC8 gene has also been implicated in insulin response in Mexican-American subjects [103] and also in type II diabetes in French Canadians [104].

3.12   ABCD1 AND X-LINKED ADRENOLEUKODYSTROPHY

The ABCD1 (ALD) gene maps to Xq28 and expresses a half transporter which is mutated in adrenoleukodystrophy (ALD). X-ALD is an X-linked recessive disorder which is characterized by various neurodegenerative phenotypes with an onset in late childhood [105]. Adrenal deficiency is common. ALD patients have an accumulation of long unbranched saturated fatty acids, with a chain length of 25–30 carbons, in the brain and adrenal cortex. The ALD protein is located in the peroxisome, where it is believed to be involved in the transport of these very long chain fatty acids (VLCFAs).

More than 400 mutations have been documented in the ABCD1 gene. Although most of the mutations seem to be point mutations, several large intragenic deletions have also been reported [106] which also may be important.

3.13   ABCG5 AND ABCG8 IN SITOSTEROLEMIA

The ABCG5 gene maps to chromosome 2p21 and is adjacent to and arranged head-to-head with the ABCG8 gene [107–109]. ABCG5 and ABCG8 proteins are classified as half transporters. Both of these genes have been found to be mutated in families with sitosterolemia, a disease characterized by defective transport of plant and fish sterols and cholesterol [110–113]. It has been hypothesized that the two half transporters form a functional heterodimer and appear to be regulated by the same promoter [114].

Patients with mutations in either ABCG5 or ABCG8 have similarly elevated levels of sitosterol, suggesting that it is the heterodimer that is the principal transporter of sitosterol. Interestingly, Asian sitosterolemia patients have almost exclusively mutations in ABCG5, and Caucasian patients have primarily mutations in ABCG8 [115–117]. This suggests that there are independent functions of the two genes, and they could also form heterodimers to transport some noncholesterol sterols found in plants and shellfish. Recently a genome-wide association study (GWAS) has detected a single nucleotide polymorphism (D19H) in ABCG8 that is a susceptibility factor for human gallstone disease [118].

·   ·   ·   ·

CHAPTER 4

The Cystic Fibrosis Transmembrane Conductance Regulator—ABCC7

4.1   WHAT IS CYSTIC FIBROSIS?

Cystic fibrosis is an inherited (autosomal recessive) disease that can affect the lungs and the digestive system. Mutations in the CF (cystic fibrosis) gene (ABCC7) cause the production of a sticky mucous that clogs the lungs, sometimes leading to life-threatening infections and also obstruction of the pancreas preventing certain enzymes from breaking down and absorbing food into the body [119]. Around 30,000 children and adults in the US are affected by CF which extends to 70,000 worldwide. Patients displaying two severe *CFTR* alleles, such as ΔF508, typically show severe disease with inadequate secretion of pancreatic enzymes which can lead to nutritional deficiencies, bacterial infections of the lung, and obstruction of the vas deferens resulting in male infertility. Patients with at least one partially functional allele display enough residual pancreatic function to avoid the major nutritional and intestinal deficiencies. Thus, there seems to be spectrum of severity in the phenotypes caused by this gene which is inversely related with the level of CFTR activity. Clearly, other modifying genes and the environment also affect disease severity, in particular the pulmonary phenotypes [119, 120].

4.2   HOW IT ALL STARTED

The first description in a child who almost certainly died of cystic fibrosis is attributed to Professor Pieter Pauw in 1595 (Professor of Botany and Anatomy at Leiden, Netherlands); however, the documented history of this disease began in the 1930s. A Swiss pediatrician - Dr. Fanconi, wrote an early paper on the disease and called it 'celiac syndrome.' It was Dr. Dorothy Anderson who coined the term 'cystic fibrosis' in 1938 and later in the 1940s, Drs. Sidney Farber and Harry Shwachman connected the abnormal secretion of mucus to the disease [119].

4.3   THE RACE TO FIND THE CYSTIC FIBROSIS GENE

Previously, genes have been mapped to chromosomes by looking for the protein that is made. But for cystic fibrosis, the protein was then unknown, hence, there was no way to distinguish that gene from all the other ones. A new approach had to be undertaken to map this particular gene. The innovative approach that researchers made use of was to use DNA markers to find the gene's location. Just like we can pinpoint the location of particular place on a map by observing key landmarks located near our point of interest, geneticists can locate particular genes on chromosomes by figuring out which 'landmarks' or 'markers' it is in proximity with. These markers are called restriction fragment-length polymorphisms or RFLPs. They were first detected in the 1970s and are placed on the chromosomes where the DNA sequence varies among individuals. These RFLPs can be used as landmarks along the chromosomes. In order to locate a gene scientists need to study the DNA of families that carry a defective gene—in this case, the cystic fibrosis gene to see if the disease trait is inherited along with any particular RFLP. If the disease is thus "linked" to the marker, the gene must be located on the same chromosome. Also, the closer the gene and the marker are on the chromosome, the less frequently they will be separated during the normal process of genetic recombination which means that there is a probability that they will be inherited together [121].

In 1985, Hans Eiberg (University Institute of Medical Genetics, Copenhagen) had discovered linkage with a gene called PON, the caveat to this being that the chromosomal location of PON was unknown, and the protein itself was very difficult to work with thus this result was not very helpful in paving out the way to the elusive CF gene. Around the same time, Lap-Chee Tsui and Manuel Buchwald (Hospital for Sick Children, Toronto) had detected linkage for a probe (a piece of single-stranded DNA that can locate particular landmarks) mapping to chromosome 7; however, due to a collaboration with Helen Donis-Keller's group at Collaborative Research Inc. (a biotech firm in Bedford, Massachusetts), they were not allowed to disclose the location of this marker immediately which generated a tremendous amount of ill will in the scientific community. However, other groups, such as Ray Whites group at the Howard Hughes Medical Institute at the University of Utah and Bob Williamsons group at St. Mary's Hospital in London had also found markers that were closer to the CF gene. Williamson's markers mapped within 10 million bp and were the T-cell receptor gene and the collagen gene while Whites group mapped it close to the MET oncogene [121]. Finally, in 1989, Lap-Chee Tsui, Jack Riordan, Francis Collins and colleagues jointly published the location of the CFTR gene on chromosome 7q31 spanning 250,000 base pairs of genomic DNA [122].

4.4   STRUCTURE OF THE CF GENE

Locus: 7q31.2—The CFTR gene is found in region q31.2 on the long (q) arm of human chromosome 7 (Figure 7).

Gene Structure: The normal allelic variant for this gene is about 250,000 base pairs long and contains 27 exons.

mRNA: The intron-free mRNA transcript for the CFTR gene is 6128 bp long. It has 6 alternative splice variants.

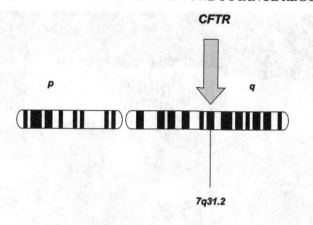

Chromosome 7

FIGURE 7: Location of the human ABCC7 (CFTR) gene. The CFTR gene is located on chromosome 7, position 7q31.2 (yellow arrow).

4.5   ABCC7—STRUCTURE, FUNCTION, AND ROLE IN CYSTIC FIBROSIS

The cystic fibrosis gene on chromosome 7q31 codes for a protein called the cystic fibrosis transmembrane conductance regulator (CFTR) also known as ABCC7. CFTR is an atypical transporter since it is an ion channel. Anion flow through the channel is required for proper functioning of the epithelia in various locations in the human body such as the cells that line the airways in the lung and also the intestinal tract. CF epithelia behave as if they are impermeable to chloride. This leads to often fatal conditions in CF patients. Mortality in CF is largely due to chronic bacterial infection and inflammation in the lung due to diminished salt absorption that reduces the killing activity of antibacterial substances on the airway surface, resulting in the recurrent colonization by opportunistic microorganisms.

It has been found that over two thirds of all CF case is caused by a single mutation in CFTR—the deletion of Phenyl Alanine 508 in NBD1 which would have important functional consequences for the protein. CFTR protein with this particular mutation fails to mature properly and is then tagged for degradation and finally becomes degraded.

To date, there are over 1000 CF disease-associated mutations [123].

CFTR is a 1480-amino acid protein with a molecular mass of 17,000 kD and has a unique structural arrangement with the overall domain organization being: TMD1-NBD1-R-TMD2-NBD2. The R is the regulatory domain which is characteristic only of CFTR. The chloride ion

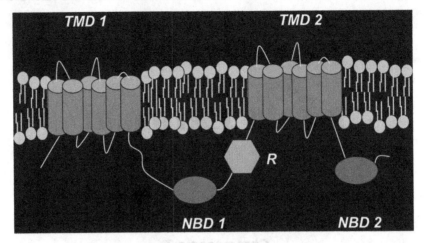

FIGURE 8: Cartoon of CFTR indicating its distinguishing features. The CFTR protein has in addition to the 2 TMD's and 2 NBD's an additional 'R' or regulatory domain.

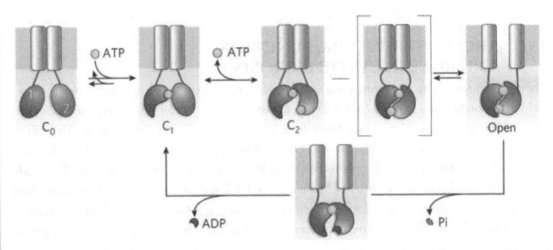

FIGURE 9: Present structural interpretation of ATP-dependent gating cycle of phosphorylated CFTR channels. The R domain is omitted. ATP (yellow) remains tightly bound to NBD1 (green) Walker motifs for several minutes, during which time many closed–open– closed gating cycles occur. ATP binding to NBD2 (blue) is followed by a slow channel opening step (C2-to-Open) that proceeds through a transition state (square brackets) in which the intramolecular NBD1–NBD2 tight heterodimer is formed but the transmembrane pore (grey rectangles) has not yet opened. The relatively stable open state becomes destabilized by hydrolysis of the ATP bound at the NBD2 composite catalytic site and loss of the hydrolysis product, inorganic phosphate (Pi). The ensuing disruption of the tight dimer interface leads to channel closure. Reproduced with permission from Nature Publishing Group, Gadsby DC, Vergani P, Csanady L. *Nature*. 2006 Mar 23; 440(7083):477–83 [124].

gate is probably opened when ATP binds to the CFTR NBD's and brings both of them into close proximity (Figure 8 and Figure 9). Hydrolysis of one of the ATP disrupts the binding together of the NBDs that can then close the channel gate and stop ion flow (Figure 9). Kinase-mediated phosphorylation of the cytoplasmic R domain is required for the transmission of the signal from the NBDs to the channel gate [124]. The R domain contains a large number of charged residues and consensus sequences for phosphorylation.

4.6 DETAILED STRUCTURE OF THE CFTR CHLORIDE CHANNEL PROTEIN

4.6.1 Transmembrane Domains

The transmembrane domains M1–M12 make up around 19% of the CFTR protein. Each TMD is made up of 6 transmembrane alpha helices, which can come together to form the pore. Many of the residues in the TMDs play a role in the formation of the CFTR channel pore.

4.6.2 The CFTR Channel Pore

Extensive mutation studies have not been carried out in CFTR to determine the amino acids that line the pore. The focus has been mainly on the 6th membrane-spanning helix of TMD1 because its cytoplasmic extension connects TMD1 to NBD1 [124]. Arg334 and Lys 335 are thought to promote entry of anions into the pore, and Thr338 seems to face the pore.

4.6.3 The R Domain

The R domain is located between NBD1 and TMD2 and contains several potential sites for phosphorylation by cAMP-dependent PKA or PKC. Dulhanty and Riordan have put forward a model of the R-domain which has two distinct domains, residues 587 to 672 (called R1) and 679 to 798 (R2). The N terminal portion of the R domain (RD1) is somewhat conserved between species; however, there is a lower degree of conservation of RD2. As mentioned before, kinase-mediated phosphorylation of the cytoplasmic R domain is required for the transmission of the signal from the NBDs to the channel gate [124]. Therefore, the main function of the R domain is in gating of the CFTR channel.

4.6.4 The NBDs

The NBDs contain a number of highly conserved motifs predicted to bind and hydrolyze ATP. The NBDs of CFTR conform to the established architecture of the typical NBD containing a Walker A motif separated from the Walker B by the C-loop or the ABC signature sequence containing the motif LSGGQ. Site-directed mutagenesis at these motifs have indicated that ATP binds to both NBDs to control the gating of the channel.

There is around 27% homology between NBDs 1 and 2.

4.6.5   Extracellular and Intracellular Domains

Approximately 4% of the CFTR protein is found in the extracellular loops. The loops are designated according to the membrane-spanning regions they connect, M1–M2, M3–M4, M5–M6, M7–M8, M9–M10, and M11–M12. CFTR also has a number of intracellular loops, ICLs. Sequences within the intracellular loops (ICL1–4) have been shown to be important for the processing of CFTR and the correct delivery of the protein to the cell membrane.

4.7   THERE IS STILL HOPE: THERAPEUTIC INTERVENTION IN CYSTIC FIBROSIS

The treatment for CF depends upon the stage of the disease and which organs are involved. Many pharmacological agents and therapeutic interventions are available to improve the lives of most people with cystic fibrosis, including antibiotics, physical therapy, exercise, anti-inflammatories, and bronchodilators to open up the airways, etc. These interventions help CF patients to lead normal and productive lives. Gene therapy also is being researched as a possible therapeutic intervention for CF.

. . . .

CHAPTER 5

PXE

5.1 WHAT IS PXE?

PXE or *Pseudoxanthoma elasticum* is a genetic disorder that affects multiple systems. The disease is characterized by mineralization of the soft connective tissue which mainly affects the skin, eyes and the arterial blood vessels [125] (Figure 10). It is designated as a 'pseudoxanthoma' since its clinical identity is separate from that of xanthomas (a xanthoma is a clinical condition in which certain fats build up underneath the skin). PXE is inherited as an autosomal recessive disease. It is estimated to affect anywhere between 1 in 25,000 and 1 in 100,000 people. Women are twice as more likely to be affected than men, and while the disease occurs in all ethnicities, it is highly prevalent in Afrikaaners due to the presence of a 'founder effect' (the mutation was present in the small group of people from whom the Afrikaaners originated). While there is no known treatment effective at combating PXE, it has been proposed that it is caused primarily by mutations in the *ABCC6* gene. PXE is a disease with high phenotypic variability and over 100 documented mutations. This gene encodes the transmembrane protein ABCC6, which may function as a transporter and is mainly expressed in the kidneys and liver.

5.2 DISCOVERY AND STRUCTURE OF THE PXE GENE

The PXE gene was discovered in 2000 by PXE International and members of the PXE International Research Consortium (PIRC). The main players in the discovery were Drs. Jouni Uitto, Charles Boyd, and Arthur Bergen, they localized the gene to the short arm of chromosome 16. The PXE gene had been previously described as the *MRP6* gene and is now designated as the *ABCC6* gene. The specific location of this gene is 16p13.1 (Figure 11), and it encompasses approximately 75 kb of DNA and is comprised of 31 exons and 32 introns. The mRNA encodes a polypeptide of 1503 aa's. There are also 2 pseudogenes related to the ABCC6 gene upstream of ABCC6 in the genome. The first pseudogene, which is designated ABCC6-Ψ1, includes the upstream region of the gene plus homologous sequence of exon 1 through intron 9. The second pseudogene, designated ABCC6Ψ2, includes the upstream sequence and the region between exon 1 and intron 4 [125–127]. There is 99% sequence similarity between ABCC6 and the pseudogenes. The ABCC6

Classic PXE   PXE-like

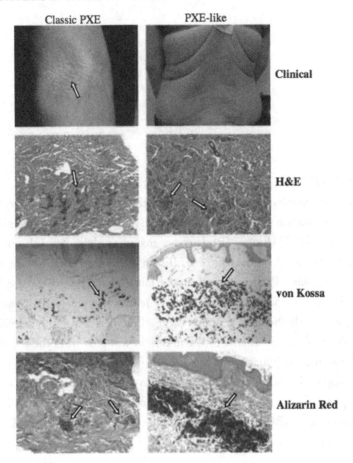

Clinical

H&E

von Kossa

Alizarin Red

FIGURE 10: Features of the skin in a PXE patient. Clinical and histopathologic features of a patient with the classic form of PXE (left panel) as well as in patients with PXE-like cutaneous features and coagulation deficiency (right panel). Reproduced with permission from Wiley, Li Q, Jiang Q, Pfendner E, Varadi A, Uitto J. *Exp Dermatol.* 2009 Jan;18(1):1–11, [125].

gene is primarily expressed in the liver, proximal tubules of the kidney, and at a very low level (if at all) in PXE-affected tissues of patients.

5.3   ABCC6—STRUCTURE, FUNCTION, AND ROLE IN PXE
5.3.1   Structural Organization of the ABCC6 Protein
The ABCC6 protein contains 3 TMDs—TMD0, TMD1, and TMD2. TMD0 is predicted to consist of 5 transmembrane alpha helices, while the other TMDs contain six, giving a total of 17 total

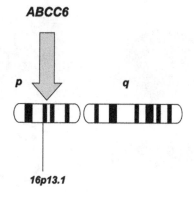

Chromosome 16

FIGURE 11: Location of the human ABCC6 gene. The human ABCC6 gene is located on chromosome 16, position 16p13.1 (yellow arrow).

transmembrane helices. TMD1 and TMD2 are postulated to be the sites for substrate binding. The role of TMD0 is unclear. The NBD organization is similar to the typical organization of most ABC transporters containing a Walker A, Walker B, C-loop, and other structures.

5.3.2   Predicted Functional Role of ABCC6 in PXE

The physiological role of the ABCC6 protein in the body is currently unknown, also its in vivo physiological substrate is yet to be discovered [125].

There are currently 2 hypotheses postulating the role of ABCC6 in PXE:

1. The Metabolic Hypothesis
2. The PXE Cell Hypothesis

5.3.2.1 The Metabolic Hypothesis. This theory puts forward the hypothesis that the lack of ABCC6 functional activity mainly in the liver results in the deficiency of certain kinds of circulatory factors which are required to prevent abnormal mineralization in the peripheral tissues (Figure 12). There is some support for this theory in studies involving the ABCC6 knockout mouse model [128] and also from various in-vitro studies [129].

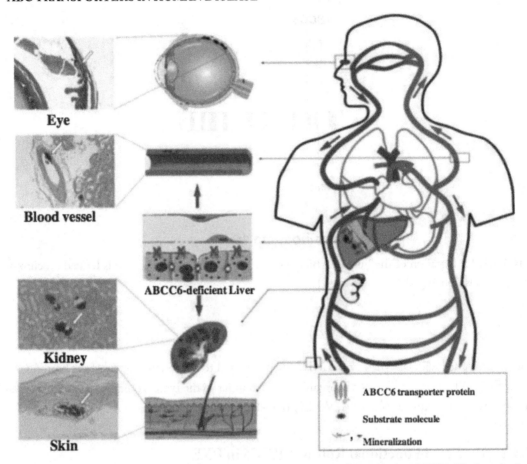

FIGURE 12: Conceptual illustration of the proposed metabolic hypothesis of PXE. Under physiologic conditions, the ABCC6 protein is expressed in high levels in the liver, presumably transporting critical metabolites to the circulation (right panel). In the absence of ABCC6 transporter activity in the liver, changes in the concentration of such substrate molecules in the circulation can take place, and the changes result in mineralization of a number of tissues, such as the eye, the arterial blood vessels, the kidney and the skin (middle panel). The presence of mineralization is detected in transgenic Abcc6(–/–) mice that recapitulate features of human PXE, by Alizarin Red stain. Reproduced with permission from Wiley, Li Q, Jiang Q, Pfendner E, Varadi A, Uitto J. *Exp Dermatol.* 2009 Jan;18(1):1–11 [125].

5.3.2.2 The PXE Cell Hypothesis. This hypothesis postulates that the absence of ABCC6 expression in PXE-affected tissues changes their biosynthetic expression profile and cell–extracellular matrix interactions which gives rise to alterations in their proliferative capacity. One line of evidence in support of this theory is that it has been found in cultured skin fibroblasts from PXE patients that MMP-2 activity is enhanced which may lead to an enhanced potential for degradation [130].

5.3.3 Speculative Working Model for the Role of ABCC6 and Matrix Gla Protein in PXE

A number of modifier genes/proteins have been proposed to influence the PXE phenotype among these proteins are MGP (Matrix Gla Protein) and GGCX (Gamma glutamyl carboxylase). The hypothesis as it stands is: if ABCC6 transports vitamin K then the mutated ABCC6 protein would be unable to transport vitamin K. Vitamin K is required as a co-factor for the carboxylation of MGP along with GGCX. As a result, MGP would be undercarboxylated which would result in the pathological mineralization of connective tissue which could have been kept at bay where MGP fully carboxylated [126].

5.4   MUTATIONS IN ABCC6 ASSOCIATED WITH PXE

Figures 13a and 13b depict the mutations in the ABCC6 gene that have been associated with the PXE phenotype. Two mutations are commonly found in Caucasians, one of them is R1141X in exon 24 which is around 30% of all PXE mutations. The second one is a AluI-mediated deletions of exons 23 to 29 (del 23–29), this has been found in at least 20% of US PXE patients (at least one allele). Other recurrent mutations are: nonsense mutations (a codon is changed to a stop codon due to introduction of a mutation) Q378X in exon 9, R518X in exon 12, R1164X in exon 24, and a clustering of nonsense mutations in exons 24 and 28 which are present in the NBDs that are critical for ABCC6 function [125–127].

The positions of nonsense, splice junction, insertion and deletion mutations identified in the ABCC6 gene in patients with pseudoxanthoma elasticum (PXE).

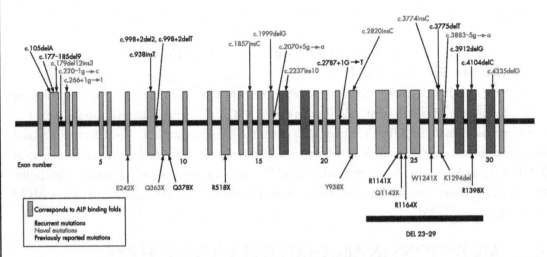

FIGURE 13A: The positions of nonsense, splice junction, insertion and deletion mutations identified in the ABCC6 gene in patients with pseudoxanthoma elasticum (PXE). Vertical blue boxes represent the 31 exons, and every fifth exon is numbered. Splicing, small insertion and deletion mutations are shown above the line, and nonsense mutations below, with the position of the recurrent del23–29 mutation. Dark blue boxes, nucleotide-binding fold domains; bold, recurrent mutations; red, novel mutations. Reproduced with permission from BMJ Publishing group, Pfendner et al. *J Med Genet* 2007;44:621–628 [127].

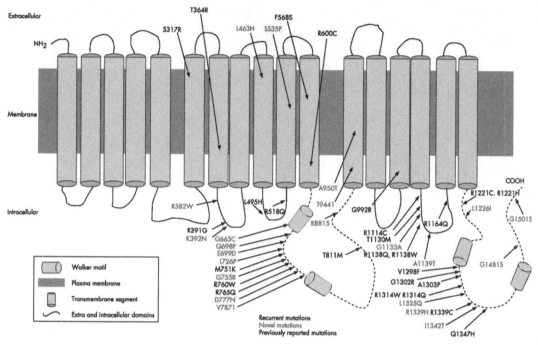

FIGURE 13B: Schematic representation of the MRP6 protein and the positions of missense mutations identified in patients with pseudoxanthoma elasticum (PXE). Vertical blocks are the 17 transmembrane domains of the MRP6 protein; dotted lines, nucleotide-binding fold; bold, recurrent missense mutations; red, novel mutations. Reproduced with permission from BMJ Publishing group, Pfendner et al. *J Med Genet* 2007;44:621–628 [127].

5.5   THERAPEUTIC INTERVENTION IN PXE

Currently, there is no therapy that interferes with the progression of PXE restriction of calcium intake and increase in vitamin K intake have been attempted but have met with limited success. To treat the manifestations of the disease plastic surgery in most cases can aleviate loose or sagging skin, laser therapy, and antiangiogenic drugs are used to treat the abnormal growth of retinal blood vessels [126].

CHAPTER 6

X-linked Adrenoleukodystrophy

6.1   WHAT IS X-ALD?

ALD or Adrenoleukodystrophy is an X-linked disorder which results in an apparent defect in peroxisomal beta oxidation and the accumulation of unbranched saturated very long chain fatty acids (VLCFA) in all tissues of the body. VLCFAs have hydrocarbon chains ranging from 25 to 30 residues in length. The symptoms of the disorder occur primarily in the adrenal cortex, the myelin of the central nervous system, and the Leydig cells of the testes. ALD belongs to a group of disorders called 'leukodystrophies,' the characteristic feature of which is damage to myelin. The age of onset for childhood ALD is usually variable (4–10 years) as are the symptoms. The myelin sheath is gradually stripped away leading to progressive loss of body function and loss of the ability to talk, this is coupled with the accumulation of VLCFs in the body tissues that cannot be disposed of. In a typical case of ALD, there is progressive brain damage and failure of the adrenal glands eventually leading to death. Mutation(s) in the ABC transporter ABCD1 also known as ALDP is thought to be a major causative factor in this disease. The incidence of X-ALD ranges between 1 in 20,000 and 1 in 100,000 and does not seem to have an ethnic predilection. The disease was brought into focus in the 1992 movie Lorenzo's Oil which is based on the true story of Augusto and Michaela Odone, the parents of Lorenzo Odone who search for a cure for their son's ALD and in the process help to invent a concoction called Lorenzo's Oil (a combination of erucic and oleic acid) [131].

6.2   LOCATION AND STRUCTURE OF THE ABCD1 GENE

The ABCD1 gene was isolated by positional cloning (positional cloning involves the isolation of partially overlapping DNA segments from genomic libraries to progress along the chromosome toward a specific gene) and is located on the X-chromosome at Xq28 (Figure 14). It is linked very closely to the cluster of colorblindness genes on the X-chromosome. There are 3 alternative transcripts of this gene. The largest one is 3664 bp containing 10 exons and spans 20kb of genomic DNA. This transcript gives rise to a protein containing 745 aa's which is localized to the peroxisomal membrane.

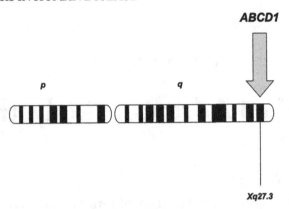

FIGURE 14: Location of the human ABCD1 gene. ABCD1 gene is located on the X-chromosome, position Xq27.3 (yellow arrow).

6.3   ABCD1—FUNCTIONAL ROLE IN ALD

ABCD1 is a peroxisomal half transporter and has been proposed to have a role in the transport of VLCFAs or their co-enzyme A derivatives into peroxisomes. All X-ALD patients have a mutation(s) in ABCD1, suggesting that the defective ABCD1 cannot transport VLCFA into the peroxisomes for degradation hence the VLCFAs start to accumulate in the cells; however, the exact mechanisms that link the VLCFA excess to inflammation and demyelination in X-ALD remain elusive. Immunocytochemical studies have shown that 70% of X-ALD patients lack the ABCD1 protein.

In the wake of recent evidence, Singh et al. [132] proposed a three-hit hypothesis for the disease pathogenesis of X-ALD. Their hypothesis states that metabolic derangements characterized by excess of VLCFA and lower plasmalogen (vinyl ether lipids with anti-oxidant ability) levels and oxidative stress (first hit) lead to inflammatory disease (second hit) with the participation of environmental, stochastic, genetic, or epigenetic factors. Subsequently, inflammatory response further causes a generalized loss of peroxisomes/peroxisomal functions (third hit), which would result in cell loss and progressive inflammatory demyelinating disease which is the characteristic feature of X-ALD.

6.3.1   Structure of the ABCD1 Protein

The ABCD1 protein is a half-transporter (Figure 15) composed of 745 amino acids. It contains 5–6 transmembrane alpha helices and one NBD. The NBD contains the traditional Walker A, Walker B, and C-loop segments. The peroxisomal half transporters contain two additional conserved motifs—the first being an EAA-like motif comprising 15 aa's which is present between TMSs 4 and 5,

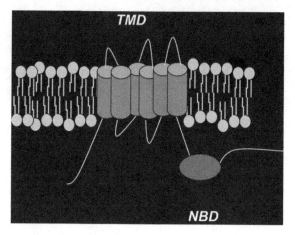

FIGURE 15: Schematic representation of the ABCD1 protein. ABCD1 protein posesses one trans-membrane domain (TMD, orange) and one nucleotide binding domain (NBD, red).

this resembles the 30aa EAA motif found in prokaryotic ABC transporters. The second conserved motif is the NPDQ motif which is located between TMS's 2 and 3 [10].

6.4   MUTATIONS IN ABCD1 ASSOCIATED WITH X-ALD

In total, there have been 1236 mutations described in X-ALD, out of which 582 are non-recurrent. Although most mutations in ABCD1 are point mutations, several large intragenic deletions have also been described [133]. Seventy-four percent of mutations in ABCD1 actually result in the absence of ALDP.

6.5   THERAPIES FOR X-ALD

A treatment consisting of erucic acid, a C22 monounsaturated fat, and oleic acid, a C18 monoun-saturated fat (Lorenzo's oil), was developed which does in some cases result in a normalization of the VLCFA levels in the blood of patients but does not appear to dramatically slow the progression of the disease [134]. This is probably because the treatment fails to lower fatty acid levels in the brain [135]. Corticosteroid treatment is recommended for those patients with adrenal insufficiency.

•   •   •   •

CHAPTER 7

ABC Proteins: A Global Perspective

ABC proteins, as we have learned, have highly conserved sequence homology and, in most cases, display similar functions across both the prokaryotic and eukaryotic realms of life. The discovery of these elaborately complex systems dates back to the 1960s and the study of bacterial transport systems. Subsequently, it was found that these bacterial transport systems were driven by harnessing the power of ATP and hence were energy-dependent systems [136]. From that point onwards, there was no looking back and, to date, these energy driven pumps have been found to be ubiquitous in distribution and functionally diverse. We have explored in these pages the role of these transporters in human disease; however, there is still much to be discovered and much more to learn about these elaborately complex yet elusively simple proteins. Some of the questions that still remain to be addressed are:

- Is there a common mechanistic principal for the functioning of these transporters by which they can recognize diverse substrates?
- How does the energy released by ATP hydrolysis facilitate the opening of the transport pore?
- and finally how are molecules actually transported across the membrane? [10].

Through three billion years of evolution, these molecular machines have diversified into essential transport systems and in the process have brought together, in recent times, a group of multidisciplinary scientists across the world who have devoted their lives to the study of these molecular motors for the betterment of human beings across the globe.

References

[1]   Jones, P.M., and George, A.M. The ABC transporter structure and mechanism: perspectives on recent research. *Cell Mol Life Sci* 61(6): pp. 682–99, 2004.

[2]   Higgins, C.F. ABC transporters: from microorganisms to man. *Ann Rev Cell Biol* 8: pp. 67–113, 1992.

[3]   Hughes, A.L. Evolution of the ATP-binding-cassette transmembrane transporters of vertebrates. *Mol Biol Evol* 11: pp. 899–910, 1994.

[4]   Allikmets, R., Gerrard, B., Hutchinson, A., and Dean, M. Characterization of the human ABC superfamily: isolation and mapping of 21 new genes using the expressed sequence tags database. *Hum Mol Genet* 5: pp. 1649–55, 1996.

[5]   Hyde, S.C., Emsley, P., Hartshorn, M.J., Mimmack, M.M., Gileadi, U., Pearce, S.R., Gallagher, M.P., Gill, D.R., Hubbard, R.E., Higgins, C.F. Structural model of ATP-binding proteins associated with cystic fibrosis, multidrug resistance and bacterial transport. *Nature* 346: pp. 362–5, 1990.

[6]   Dean, M., Rzhetsky, A., Allikmets, R. The human ATP-binding cassette (ABC) transporter superfamily. *Genome Res* 11: pp. 1156–66, 2001.

[7]   Higgins, C.F. ABC transporters: physiology, structure and mechanism—an overview. *Res Microbiol* 152: pp. 205–10, 2001.

[8]   Moitra, K., Lou, H., and Dean, M. Multidrug efflux pumps and cancer stem cells: insights into multidrug resistance and therapeutic development. *Clin Pharmacol Ther* 89(4): pp. 491–502, 2011.

[9]   Aller, S.G., Yu, J., Ward, A., Weng, Y., Chittaboina, S., Zhuo, R., Harrell, P.M., Trinh, Y.T., Zhang, Q., Ina, L., Urbatsch, I.L., Chang, G. Structure of p-glycoprotein reveals a molecular basis for poly-specific drug binding. *Science* 323: pp. 1718–22, 2009.

[10]   ABC Proteins from Bacteria to Man. Elsevier Sciences Ltd. Editors: Holland, B.I., Cole, S.P.C., Kuchler, K., and Higgins, C.F. ISBN 0-12-352551-9, 2003.

[11]   Venter, J.C., Adams, M.D., Myers, E.W., Li, P.W., Mural, R.J., Sutton, G.G., Smith, H.O., Yandell, M., Evans, C.A., Holt, R.A., et al. The sequence of the human genome. *Science* 291: pp. 1304–51, 2001.

[12]   Lander, E.S., Linton, L.M., Birren, B., Nusbaum, C., Zody, M.C., Baldwin, J., Devon, K.,

Dewar, K., Doyle, M., FitzHugh, W., et al. Initial sequencing and analysis of the human genome. International Human Genome Sequencing Consortium. *Nature* 409: pp. 860–921, 2001.

[13]   The Human ATP-Binding Cassette (ABC) Transporter Superfamily [Internet]. Dean M. Bethesda (MD): National Center for Biotechnology Information (US), 2002.

[14]   Broccardo, C., Luciani, M., Chimini, G. The ABCA subclass of mammalian transporters. *Biochim Biophys Acta* 1461: pp. 395–404, 1999.

[15]   Arnould, I., Schriml, L., Prades, C., Lachtermacher-Triunfol, M., Schneider, T., Maintoux, C., Lemoine, C., Debono, D., Devaud, C., Naudin, L., et al. Identification and characterization of a cluster of five new ATP-binding cassette transporter genes on human chromosome 17q24: a new sub-group within the ABCA sub-family. *GeneScreen* 1: pp. 157–64, 2001.

[16]   Anjard, C., Consortium, D.S., Loomis, W.F. Evolutionary analyses of ABC transporters of *Dictyostelium discoideum. Eukaryotic Cell* 1: pp. 643–52, 2002.

[17]   Quinton, P.M. Physiological basis of cystic fibrosis: a historical perspective. *Physiol Rev* 79: pp. S3–22, 1999.

[18]   Bakos, E., Evers, R., Calenda, G., Tusnady, G.E., Szakacs, G., Varadi, A., Sarkadi, B. Characterization of the amino-terminal regions in the human multidrug resistance protein (MRP1). *J Cell Sci* 113: pp. 4451–61, 2000.

[19]   Marton, M.J., Vazquez de Aldana, C.R., Qiu, H., Chakraburtty, K., Hinnebusch, A.G. Evidence that GCN1 and GCN20, translational regulators of GCN4, function on elongating ribosomes in activation of eIF2a kinase GCN2. *Mol Cell Biol* 17: pp. 4474–89, 1997.

[20]   Tyzack, J.K., Wang, X., Belsham, G.J., Proud, C.G. ABC50 interacts with eukaryotic initiation factor 2 and associates with the ribosome in an ATP-dependent manner. *J Biol Chem* 275: pp. 34131–9, 2000.

[21]   Chen, H., Rossier, C., Lalioti, M.D., Lynn, A., Chakravarti, A., Perrin, G., Antonarakis, S.E. Cloning of the cDNA for a human homologue of the *Drosophila white* gene and mapping to chromosome 21q22.3. *Am J Hum Genet* 59: pp. 66–75, 1996.

[22]   Klucken, J., Buchler, C., Orso, E., Kaminski, W.E., Porsch-Ozcurumez, M., Liebisch, G., Kapinsky, M., Diederich, W., Drobnik, W., Dean, M., et al. ABCG1 (ABC8), the human homolog of the *Drosophila white* gene, is a regulator of macrophage cholesterol and phospholipid transport. *Proc Natl Acad Sci U S A* 97: pp. 817–22, 2000.

[23]   Annilo, T., Chen, Z.Q., Shulenin, S., Costantino, J., Thomas, L., Lou, H., Stefanov, S., and Dean, M. Evolution of the vertebrate ABC gene family: analysis of gene birth and death. *Genomics* 88: pp. 1–11, 2006.

[24]   Ohno, S. Gene duplication and the uniqueness of vertebrate genomes circa 1970–1999. *Semin Cell Dev Biol* 10: pp. 517–22, 1999.

[25]   Zhang, J. Evolution by gene duplication: an update. *Trends Ecol Evol* 18: pp. 292–8, 2003.

[26]   Ohno, S. Evolution by gene duplication (Springer-Verlag). ISBN 0-04-575015-7, 1970.

[27]   Kimura, M. The Neutral Theory of Molecular Evolution, Cambridge University Press, Cambridge, CB2 2RU, UK. ISBN-13 978-0-521-23109-1, 1983.

[28]   Hughes, A.L., and Nei, M. Evolution of the major histocompatibility complex: independent origin of nonclassical class I genes in different groups of mammals. *Mol Biol Evol* 6: pp. 559–79, 1989.

[29]   Nei, M., Rogozin, I.B., and Piontkvska, H. Purifying selection and birth-and-death evolution in the ubiquitin gene family. *Proc Natl Acad Sci U S A* 97: pp. 10866–71, 2000.

[30]   Annilo, T., and Dean, M. Degeneration of an ATP-binding cassette transporter gene, ABCC13, in different mammalian lineages. *Genomics* 84: pp. 34–46, 2004.

[31]   Oram, J.F., Lawn, R.M. ABCA1: the gatekeeper for eliminating excess cholesterol. *J Lipid Res* 42: pp. 1173–9, 2001.

[32]   Allikmets, R., Singh, N., Sun, H., Shroyer, N.F., Hutchinson, A., Chidambaram, A., Gerrard, B., Baird, L., Stauffer, D., Peiffer, A., et al. A photoreceptor cell-specific ATP-binding transporter gene (*ABCR*) is mutated in recessive Stargardt macular dystrophy. *Nat Genet* 15: pp. 236–46, 1997.

[33]   Smith, K.D., Kemp, S., Braiterman, L.T., Lu, J.F., Wei, H.M., Geraghty, M., Stretten, G., Bergin, J.S., Pevsner, J., and Watkins, P.A. X-linked adrenoleukodystrophy: genes, mutations, and phenotypes. *Neurochem. Res* 24: pp. 521–35, 1999.

[34]   Nei, M., and Hughes, A.L. In 11th Histocompatibility Workshop and Conference, Eds. Tsuji, K., Aizawa, M., and Sasazuki, T. (Oxford Univ. Press, Oxford), Vol. 2, pp. 27–38, 1992.

[35]   Ota, T., and Nei, M. Divergent evolution and evolution by birth and death process in the immunoglobulin VH gene family. *Mol Biol Evol* 11: pp. 469–82, 1994.

[36]   Allikmets, R., Schriml, L.M., Hutchinson, A., Romano-Spica, V., Dean, M. A human placenta-specific ATP-binding cassette gene (ABCP) on chromosome 4q22 that is involved in multidrug resistance. *Cancer Res* 58: pp. 5337–9, 1998.

[37]   Doyle, L.A., Yang, W., Abruzzo, L.V., Krogmann, T., Gao, Y., Rishi, A.K., Ross, D.D. A multidrug resistance transporter from human MCF-7 breast cancer cells. *Proc Natl Acad Sci U S A* 95: pp. 15665–70, 1998.

[38]   Miyake, K., Mickley, L., Litman, T., Zhan, Z., Robey, R., Cristensen, B., Brangi, M., Greenberger, L., Dean, M., Fojo, T., et al. Molecular cloning of cDNAs which are highly overexpressed in mitoxantrone-resistant cells: demonstration of homology to ABC transport genes. *Cancer Res* 59: pp. 8–13, 1999.

[39]   Ross, D.D., Yang, W., Abruzzo, L.V., Dalton, W.S., Schneider, E., Lage, H., Dietel, M.,

Greenberger, L., Cole, S.P., Doyle, L.A. Atypical multidrug resistance: breast cancer resistance protein messenger RNA expression in mitoxantrone-selected cell lines. *J Natl Cancer Inst* 91: pp. 429–33, 1999.

[40] Mickley, L., Jain, P., Miyake, K., Schriml, L.M., Rao, K., Fojo, T., Bates, S., Dean, M. An ATP-binding cassette gene (*ABCG3*) closely related to the multidrug transporter ABCG2 (MXR/ABCP) has an unusual ATP-binding domain. *Mamm Genome* 12: pp. 86–8, 2001.

[41] Roninson, I.B., Chin, J.E., Choi, K., Gros, P., Housman, D.E., Fojo, A., Shen, D.W., Gottesman, M.M., Pastan, I. Isolation of human mdr DNA sequences amplified in multidrug-resistant KB carcinoma cells. *Proc Natl Acad Sci U S A* 83: pp. 4538–42, 1986.

[42] Ruetz, S., Gros, P. Phosphatidylcholine translocase: a physiological role for the *mdr2* gene. *Cell* 77: pp. 1071–81, 1994.

[43] Van Helvoort, A., Smith, A.J., Sprong, H., Fritzsche, I., Schinkel, A.H., Borst, P., van Meer, G. MDR1 P-glycoprotein is a lipid translocase of broad specificity, while MDR3 P-glycoprotein specifically translocates phosphatidylcholine. *Cell* 87: pp. 507–17, 1996.

[44] Shulenin, S., Schriml, L.M., Remaley, A.T., Fojo, S., Brewer, B., Allikmets, R., Dean, M. An ATP-binding cassette gene (ABCG5) from the *ABCG* (White) gene subfamily maps to human chromosome 2p21 in the region of the sitosterolemia locus. *Cytogenet Cell Genet* 92: pp. 204–8, 2001.

[45] Lee, M.H., Lu, K., Hazard, S., Yu, H., Shulenin, S., Hidaka, H., Kojima, H., Allikmets, R., Sakuma, N., Pegoraro, R., et al. Identification of a gene, *ABCG5*, important in the regulation of dietary cholesterol absorption. *Nat Genet* 27: pp. 79–83, 2001.

[46] Remaley, A.T., Bark, S., Walts, A.D., Freeman, L., Shulenin, S., Annilo, T., Elgin, E., Rhodes, H.E., Joyce, C., Dean, M., et al. Comparative genome analysis of potential regulatory elements in the *ABCG5–ABCG8* gene cluster. *Biochem Biophys Res Commun* 295: pp. 276–82, 2002.

[47] Berge, K.E., Tian, H., Graf, G.A., Yu, L., Grishin, N.V., Schultz, J., Kwiterovich, P., Shan, B., Barnes, R., Hobbs, H.H. Accumulation of dietary cholesterol in sitosterolemia caused by mutations in adjacent ABC transporters. *Science* 290: pp. 1771–5, 2000.

[48] Dean, M., and Annilo, T. Evolution of the ATP-binding cassette (ABC) transporter superfamily in vertebrates. *Ann Rev Genomics Hum Genet* 6: pp. 123–42, 2005.

[49] Ban, N., Sasaki, M., Sakai, H., Ueda, K., Inagaki, N. Cloning of ABCA17, a novel rodent sperm-specific ABC (ATP-binding cassette) transporter that regulates intracellular lipid metabolism. *Biochem J* 389(2): pp. 577–85, 2005.

[50] Klein, J., Sato, A., and O'hugin, C. Evolution by gene duplication in the major histocompatibility complex. *Cytogenet Cell Genet* 80: pp. 123–7, 1998.

[51] Klein, I., Sarkadi, B., Varadi, A. An inventory of the human ABC proteins. *Biochim Biophys Acta* 1461: pp. 237–62, 1999.

[52]   Bodzioch, M., Orso, E., Klucken, J., Langmann, T., Bottcher, A., Diederich, W., Drobnik, W., Barlage, S., Buchler, C., Porsch-Ozcurumez, M., et al. The gene encoding ATP-binding cassette transporter 1 is mutated in Tangier disease. *Nat Genet* 22: pp. 347–51, 1999.

[53]   Brooks-Wilson, A., Marcil, M., Clee, S.M., Zhang, L.H., Roomp, K., van Dam, M., Yu, L., Brewer, C., Collins, J.A., Molhuizen, H.O., et al. Mutations in ABC1 in Tangier disease and familial high-density lipoprotein deficiency. *Nat Genet* 22: pp. 336–45, 1999.

[54]   Rust, S., Rosier, M., Funke, H., Real, J., Amoura, Z., Piette, J.C., Deleuze, J.F., Brewer, H.B., Duverger, N., Denefle, P., et al. Tangier disease is caused by mutations in the gene encoding ATP-binding cassette transporter 1. *Nat Genet* 22: pp. 352–5, 1999.

[55]   Remaley, A.T., Rust, S., Rosier, M., Knapper, C., Naudin, L., Broccardo, C., Peterson, K.M., Koch, C., Arnould, I., Prades, C., et al. Human ATP-binding cassette transporter 1 (ABC1): genomic organization and identification of the genetic defect in the original Tangier disease kindred. *Proc Natl Acad Sci U S A* 96: pp. 12685–90, 1999.

[56]   Young, S.G., Fielding, C.J. The ABCs of cholesterol efflux. *Nat Genet* 22: pp. 316–8, 1999.

[57]   Allikmets, R., Gerrard, B., Hutchinson, A., Dean, M. Characterization of the human ABC superfamily: isolation and mapping of 21 new genes using the expressed sequence tags database. *Hum Mol Genet* 5: pp. 1649–55, 1996.

[58]   Allikmets, R., Singh, N., Sun, H., Shroyer, N.F., Hutchinson, A., Chidambaram, A., Gerrard, B., Baird, L., Stauffer, D., Peiffer, A., et al. A photoreceptor cell-specific ATP-binding transporter gene *(ABCR)* is mutated in recessive Stargardt macular dystrophy. *Nat Genet* 15: pp. 236–46, 1997.

[59]   Azarian, S.M., Travis, G.H. The photoreceptor rim protein is an ABC transporter encoded by the gene for recessive Stargardt's disease (ABCR). *FEBS Lett* 409: pp. 247–52, 1997.

[60]   Allikmets, R. Simple and complex ABCR: genetic predisposition to retinal disease. *Am J Hum Genet* 67: pp. 793–9, 2000.

[61]   Martinez-Mir, A., Paloma, E., Allikmets, R., Ayuso, C., del Rio, T., Dean, M., Vilageliu, L., Gonzalez-Duarte, R., Balcells, S. Retinitis pigmentosa caused by a homozygous mutation in the Stargardt disease gene ABCR [letter; comment]. *Nat Genet* 18: pp. 11–2, 1998.

[62]   Rozet, J.M., Gerber, S., Ghazi, I., Perrault, I., Ducroq, D., Souied, E., Cabot, A., Dufier, J.L., Munnich, A., Kaplan, J. Mutations of the retinal specific ATP binding transporter gene (ABCR) in a single family segregating both autosomal recessive retinitis pigmentosa RP19 and Stargardt disease: evidence of clinical heterogeneity at this locus. *J Med Genet* 36: pp. 447–51, 1999.

[63]   Cremers, F.P., van de Pol, D.J., van Driel, M., den Hollander, A.I., van Haren, F.J., Knoers, N.V., Tijmes, N., Bergen, A.A., Rohrschneider, K., Blankenagel, A., et al. Autosomal

recessive retinitis pigmentosa and cone-rod dystrophy caused by splice site mutations in the Stargardt's disease gene *ABCR*. *Hum Mol Genet* 7: pp. 355–62, 1998.

[64]   Stargardt, K. Uber familiare, progressive degeneration in der maculagegend des auges. *Albrecht van Graefes Arch Ophthalmol* 71: pp. 534–50, 1909.

[65]   Molday, R.S., Zhong, M., Quazi, F. The role of the photoreceptor ABC transporter ABCA4 in lipid transport and Stargardt macular degeneration. *Biochim Biophys Acta* 1791(7): pp. 573–83, 2009.

[66]   Allikmets, R., Shroyer, N.F., Singh, N., Seddon, J.M., Lewis, R.A., Bernstein, P.S., Peiffer, A., Zabriskie, N.A., Li, Y., Hutchinson, A., et al. Mutation of the Stargardt disease gene *(ABCR)* in age-related macular degeneration. *Science* 277: pp. 1805–7, 1997.

[67]   Juliano, R.L., Ling, V.A. A surface glycoprotein modulating drug permeability in Chinese hamster ovary cell mutants. *Biochim Biophys Acta* 455: pp. 152–62, 1976.

[68]   Riordan, J.R., Deuchars, K., Kartner, N., Alon, N., Trent, J., Ling, V. Amplification of P-glycoprotein genes in multidrug-resistant mammalian cell lines. *Nature* 316: pp. 817–9, 1985.

[69]   Kartner, N., Evernden-Porelle, D., Bradley, G., Ling, V. Detection of P-glycoprotein in multidrug-resistant cell lines by monoclonal antibodies. *Nature* 316: pp. 820–3, 1985.

[70]   Roninson, I.B., Chin, J.E., Choi, K., Gros, P., Housman, D.E., Fojo, A., Shen, D.-W., Gottesman, M.M., Pastan, I. Isolation of human mdr DNA sequences amplified in multidrug-resistant KB carcinoma cells. *Proc Natl Acad Sci U S A* 83: pp. 4538–42, 1986.

[71]   Schinkel, A.H., Mayer, U., Wagenaar, E., Mol, C.A., van Deemter, L., Smit, J.J., van der Valk, M.A., Voordouw, A.C., Spits, H., van Tellingen, O., et al. Normal viability and altered pharmacokinetics in mice lacking mdr1-type (drug-transporting) P-glycoproteins. *Proc Natl Acad Sci U S A* 94: pp. 4028–33, 1997.

[72]   Chaudhary, P.M., Roninson, I.B. Expression and activity of P-glycoprotein, a multidrug efflux pump, in human hematopoietic stem cells. *Cell* 66: pp. 85–94, 1991.

[73]   Kimchi-Sarfaty, C., Oh, J.M., Kim, I.W., Sauna, Z.E., Calcagno, A.M., Ambudkar, S.V., Gottesman, M.M. A "silent" polymorphism in the MDR1 gene changes substrate specificity. *Science* 315(5811): pp. 525–8, 2007.

[74]   Trowsdale, J., Hanson, I., Mockridge, I., Beck, S., Townsend, A., Kelly, A. Sequences encoded in the class II region of the MHC related to the "ABC" superfamily of transporters [see comments]. *Nature* 348: pp. 741–4, 1990.

[75]   Spies, T., Bresnahan, M., Bahram, S., Arnold, D., Blanck, G., Mellins, E., Pious, D., DeMars, R. A gene in the human major histocompatibility complex class II region controlling the class I antigen presentation pathway. *Nature* 348: pp. 744–7, 1990.

[76]   Monaco, J.J., Cho, S., Attaya, M. Transport protein genes in the murine MHC: possible implications for antigen processing. *Science* 250: pp. 1723–6, 1990.

[77] Salter, R.D., Cresswell, P. Impaired assembly and transport of HLA-A and -B antigens in a mutant TxB cell hybrid. *EMBO J* 5: pp. 943–9, 1986.

[78] Uebel, S., Tampe, R. Specificity of the proteasome and the TAP transporter. *Curr Opin Immunol* 11: pp. 203–8, 1999.

[79] Uebel, S., Kraas, W., Kienle, S., Wiesmuller, K.H., Jung, G., Tampe, R. Recognition principle of the TAP transporter disclosed by combinatorial peptide libraries. *Proc Natl Acad Sci U S A* 94: pp. 8976–81, 1997.

[80] Herget, M., Tampé, R. Intracellular peptide transporters in human compartmentalization of the "peptidome." *Pflug Arch Eur J Phy* 453: pp. 591–600, 2007.

[81] de la Salle, H., Zimmer, J., Fricker, D., Angenieux, C., Cazenave, J.P., Okubo, M., Maeda, H., Plebani, A., Tongio, M.M., Dormoy, A., et al. HLA class I deficiencies due to mutations in subunit 1 of the peptide transporter TAP1. *J Clin Invest* 103: pp. R9–13, 1999.

[82] Lankat-Buttgereit, B., Tampé, R. The transporter associated with antigen processing: function and implications in human diseases. *Physiol Rev* 82(1): pp. 187–204, 2002.

[83] Csere, P., Lill, R., Kispal, G. Identification of a human mitochondrial ABC transporter, the functional orthologue of yeast Atm1p. *FEBS Lett* 441: pp. 266–70, 1998.

[84] Allikmets, R., Raskind, W.H., Hutchinson, A., Schueck, N.D., Dean, M., Koeller, D.M. Mutation of a putative mitochondrial iron transporter gene (ABC7) in X-linked sideroblastic anemia and ataxia (XLSA/A). *Hum Mol Genet* 8: pp. 743–9, 1999.

[85] Bekri, S., Kispal, G., Lange, H., Fitzsimons, E., Tolmie, J., Lill, R., Bishop, D.F. Human ABC7 transporter: gene structure and mutation causing X-linked sideroblastic anemia with ataxia with disruption of cytosolic iron–sulfur protein maturation. *Blood* 96: pp. 3256–64, 2000.

[86] Childs, S., Yeh, R.L., Georges, E., Ling, V. Identification of a sister gene to P-glycoprotein. *Cancer Res* 55: pp. 2029–34, 1995.

[87] Gerloff, T., Stieger, B., Hagenbuch, B., Madon, J., Landmann, L., Roth, J., Hofmann, A.F., Meier, P.J. The sister of P-glycoprotein represents the canalicular bile salt export pump of mammalian liver. *J Biol Chem* 273: pp. 10046–50, 1998.

[88] Strautnieks, S., Bull, L.N., Knisely, A.S., Kocoshis, S.A., Dahl, N., Arnell, H., Sokal, E., Dahan, K., Childs, S., Ling, V., et al. A gene encoding a liver-specific ABC transporter is mutated in progressive familial intrahepatic cholestasis. *Nat Genet* 20: pp. 233–8, 1998.

[89] Pauli-Magnus, C., Stieger, B., Meier, Y., Kullak-Ublick, G.A., Meier, P.J. Enterohepatic transport of bile salts and genetics of cholestasis. *J Hepatol* 43(2): pp. 342–57, 2005.

[90] Kool, M., de Haas, M., Scheffer, G.L., Scheper, R.J., van Eijk, M.J., Juijn, J.A., Baas, F., Borst, P. Analysis of expression of cMOAT (MRP2), MRP3, MRP4, and MRP5, homologues of the multidrug resistance-associated protein gene (*MRP1*), in human cancer cell lines. *Cancer Res* 57: pp. 3537–47, 1997.

[91]   Wada, M., Toh, S., Taniguchi, K., Nakamura, T., Uchiumi, T., Kohno, K., Yoshida, I., Kimura, A., Sakisaka, S., Adachi, Y., et al. Mutations in the canalicular multispecific organic anion transporter (*cMOAT*) gene, a novel ABC transporter, in patients with hyperbilirubinemia II/Dubin–Johnson syndrome. *Hum Mol Genet* 7: pp. 203–7, 1998.

[92]   Toh, S., Wada, M., Uchiumi, T., Inokuchi, A., Makino, Y., Horie, Y., Adachi, Y., Sakisaka, S., Kuwano, M. Genomic structure of the canalicular multispecific organic anion-transporter gene (*MRP2/cMOAT*) and mutations in the ATP-binding-cassette region in Dubin–Johnson syndrome. *Am J Hum Genet* 64: pp. 739–46, 1999.

[93]   Nies, A.T., Keppler, D. The apical conjugate efflux pump ABCC2 (MRP2). *Pflugers Arch* 453: pp. 643–59, 2007.

[94]   Cui, Y., Konig, J., Buchholz, J.K., Spring, H., Leier, I., Keppler, D. Drug resistance and ATP-dependent conjugate transport mediated by the apical multidrug resistance protein, MRP2, permanently expressed in human and canine cells. *Mol Pharmacol* 55: pp. 929–37, 1999.

[95]   Wang, J., Near, S., Young, K., Connelly, P.W., Hegele, R.A. *ABCC6* gene polymorphism associated with variation in plasma lipoproteins. *J Hum Genet* 46: pp. 699–705, 2001.

[96]   Trip, M.D., Smulders, Y.M., Wegman, J.J., Hu, X., Boer, J.M., ten Brink, J.B., Zwinderman, A.H., Kastelein, J.J., Feskens, E.J., Bergen, A.A. Frequent mutation in the *ABCC6* gene (*R1141X*) is associated with a strong increase in the prevalence of coronary artery disease. *Circulation* 106: pp. 773–5, 2002.

[97]   Riordan, J.R., Rommens, J.M., Kerem, B., Alon, N., Rozmahel, R., Grzelczak, Z., Zielenski, J., Lok, S., Plavsic, N., Chou, J.L., et al. Identification of the cystic fibrosis gene: cloning and characterization of complementary DNA. *Science* 245: pp. 1066–73, 1989.

[98]   Kerem, B., Rommens, J.M., Buchanan, J.A., Markiewicz, D., Cox, T.K., Chakravarti, A., Buchwald, M., Tsui, L.C. Identification of the cystic fibrosis gene: genetic analysis. *Science* 245: pp. 1073–80, 1989.

[99]   Rommens, J.M., Iannuzzi, M.C., Kerem, B., Drumm, M.L., Melmer, G., Dean, M., Rozmahel, R., Cole, J.L., Kennedy, D., Hidaka, N., et al. Identification of the cystic fibrosis gene: chromosome walking and jumping. *Science* 245: pp. 1059–65, 1989.

[100] Andersen, D.H. Cystic fibrosis of the pancreas and its relationship to celiac disease. *Amer J Dis Child* 56: pp. 344–99, 1938.

[101] Gabriel, S.E., Clarke, L.L., Boucher, R.C., Stutts, M.J. CFTR and outward rectifying chloride channels are distinct proteins with a regulatory relationship. *Nature* 363: pp. 263–8, 1993.

[102] Thomas, P.M., Cote, G.J., Wohllk, N., Haddad, B., Mathew, P.M., Rabl, W., Aguilar-Bryan, L., Gagel, R.F., Bryan, J. Mutations in the sulfonylurea receptor gene in familial persistent hyperinsulinemic hypoglycemia of infancy. *Science* 268: pp. 426–9, 1995.

[103]   Goksel, D.L., Fischbach, K., Duggirala, R., Mitchell, B.D., Aguilar-Bryan, L., Blangero, J., Stern, M.P., O'Connell, P. Variant in sulfonylurea receptor-1 gene is associated with high insulin concentrations in non-diabetic Mexican Americans: *SUR-1* gene variant and hyperinsulinemia. *Hum Genet* 103: pp. 280–5, 1998.

[104]   Reis, A.F., Ye, W.Z., Dubois-Laforgue, D., Bellanne-Chantelot, C., Timsit, J., Velho, G. Association of a variant in exon 31 of the sulfonylurea receptor 1 (SUR1) gene with type 2 diabetes mellitus in French Caucasians. *Hum Genet* 107: pp. 138–44, 2000.

[105]   Mosser, J., Douar, A.M., Sarde, C.O., Kioschis, P., Feil, R., Moser, H., Poustka, A.M., Mandel, J.L., Aubourg, P. Putative X-linked adrenoleukodystrophy gene shares unexpected homology with ABC transporters. *Nature* 361: pp. 726–30, 1993.

[106]   Kutsche, K., Ressler, B., Katzera, H.G., Orth, U., Gillessen-Kaesbach, G., Morlot, S., Schwinger, E., Gal, A. Characterization of breakpoint sequences of five rearrangements in *L1CAM* and *ABCD1* (*ALD*) genes. *Hum Mutat* 19: pp. 526–35, 2002.

[107]   Shulenin, S., Schriml, L.M., Remaley, A.T., Fojo, S., Brewer, B., Allikmets, R., Dean, M. An ATP-binding cassette gene (ABCG5) from the *ABCG* (White) gene subfamily maps to human chromosome 2p21 in the region of the sitosterolemia locus. *Cytogenet Cell Genet* 92: pp. 204–8, 2001.

[108]   Berge, K.E., Tian, H., Graf, G.A., Yu, L., Grishin, N.V., Schultz, J., Kwiterovich, P., Shan, B., Barnes, R., Hobbs, H.H. Accumulation of dietary cholesterol in sitosterolemia caused by mutations in adjacent ABC transporters. *Science* 290: pp. 1771–5, 2000.

[109]   Lee, M.H., Lu, K., Hazard, S., Yu, H., Shulenin, S., Hidaka, H., Kojima, H., Allikmets, R., Sakuma, N., Pegoraro, R., et al. Identification of a gene, *ABCG5*, important in the regulation of dietary cholesterol absorption. *Nat Genet* 27: pp. 79–83, 2001.

[110]   Bhattacharyya, A.K., Connor, W.E. β-Sitosterolemia and xanthomatosis. A newly described lipid storage disease in two sisters. *J Clin Invest* 53: pp. 1033–43, 1974.

[111]   Salen, G., Shefer, S., Nguyen, L., Ness, G.C., Tint, G.S., Batta, A.K. Sitosterolemia. *Subcell Biochem* 28: pp. 453–76, 1997.

[112]   Gregg, R.E., Connor, W.E., Lin, D.S., Brewer, H.B., Jr. Abnormal metabolism of shellfish sterols in a patient with sitosterolemia and xanthomatosis. *J Clin Invest* 77: pp. 1864–72, 1986.

[113]   Patel, S.B., Salen, G., Hidaka, H., Kwiterovich, P.O., Stalenhoef, A.F., Miettinen, T.A., Grundy, S.M., Lee, M.H., Rubenstein, J.S., Polymeropoulos, M.H., et al. Mapping a gene involved in regulating dietary cholesterol absorption. The sitosterolemia locus is found at chromosome 2p21. *J Clin Invest* 102: pp. 1041–4, 1998.

[114]   Remaley, A.T., Bark, S., Walts, A.D., Freeman, L., Shulenin, S., Annilo, T., Elgin, E., Rhodes, H.E., Joyce, C., Dean, M., et al. Comparative genome analysis of potential

regulatory elements in the *ABCG5–ABCG8* gene cluster. *Biochem Biophys Res Commun* 295: pp. 276–82, 2002.

[115] Lu, K., Lee, M.H., Hazard, S., Brooks-Wilson, A., Hidaka, H., Kojima, H., Ose, L., Stalenhoef, A.F., Mietinnen, T., Bjorkhem, I., et al. Two genes that map to the STSL locus cause sitosterolemia: genomic structure and spectrum of mutations involving sterolin-1 and sterolin-2, encoded by ABCG5 and ABCG8, respectively. *Am J Hum Genet* 69: pp. 278–90, 2001.

[116] Lam, C.W., Cheng, A.W., Tong, S.F., Chan, Y.W. Novel donor splice site mutation of *ABCG5* gene in sitosterolemia. *Mol Genet Metab* 75: pp. 178–80, 2002.

[117] Heimer, S., Langmann, T., Moehle, C., Mauerer, R., Dean, M., Beil, F.U., Von Bergmann, K., Schmitz, G. Mutations in the human ATP-binding cassette transporters ABCG5 and ABCG8 in sitosterolemia. *Hum Mutat* 20: p. 151, 2002.

[118] Buch, S., Schafmayer, C., Völzke, H., Becker, C., Franke, A., von Eller-Eberstein, H., et al. A genome-wide association scan identifies the hepatic cholesterol transporter ABCG8 as a susceptibility factor for human gallstone disease. *Nat Genet* 39: pp. 995–9, 2007.

[119] http://www.cff.org

[120] http://www.aboutcysticfibrosis.com

[121] Roberts, L. Race for the cystic fibrosis gene (Research News). *Science* 240: pp. 141–4, 1988.

[122] Rommens, J.M., Iannuzzi, M.C., Kerem, B., Drumm, M.L., Melmer, G., Dean, M., Rozmahel, R., Cole, J.L., Kennedy, D., Hidaka, N., et al. Identification of the cystic fibrosis gene: chromosome walking and jumping. *Science* 245(4922): pp. 1059–65, 1989.

[123] http://www.genet.sickkids.on.ca/cftr/

[124] Gadsby, D.C., Vergani, P., Csanády, L. The ABC protein turned chloride channel whose failure causes cystic fibrosis. *Nature* 440(7083): pp. 477–83, 2006.

[125] Li, Q., Jiang, Q., Pfendner, E., Váradi, A., Uitto, J. Pseudoxanthoma elasticum: clinical phenotypes, molecular genetics and putative pathomechanisms. *Exp Dermatol* 18(1): pp. 1–11, 2009.

[126] www.pxe.org

[127] Pfendner, E.G., Vanakker, O.M., Terry, S.F., Vourthis, S., McAndrew, P.E., McClain, M.R., et al. Mutation detection in the ABCC6 gene and genotype–phenotype analysis in a large international case series affected by pseudoxanthoma elasticum. *J Med Genet* 44: pp. 621–8, 2007.

[128] Klement, J.F., Matsuzaki, Y., Jiang, Q.-J., et al. Targeted ablation of the ABCC6 gene results in ectopic mineralization of connective tissues. *Mol Cell Biol* 25: pp. 8299–310, 2005.

[129] Le Saux, O., Bunda, S., VanWart, C.M., et al. Serum factors from pseudoxanthoma elas-

ticum patients alter elastic fiber formation in vitro. *J Invest Dermatol* 126: pp. 1497–505, 2006.

[130] Quaglino, D., Sartor, L., Gabrisa, S., et al. Dermal fibroblasts from pseudoxanthoma elasticum patients have raised MMP-2 degradative potential. *Biochim Biophys Acta* 1741: pp. 42–7, 2005.

[131] http://www.x-ald.nl/biochemistry-genetics/abcd-proteins/

[132] Singh, I., Pujol, A. Pathomechanisms underlying X-adrenoleukodystrophy: a three hit hypothesis. *Brain Pathol* 20: pp. 838–44, 2010.

[133] Kutsche, K., Ressler, B., Katzera, H.G., Orth, U., Gillessen-Kaesbach, G., Morlot, S., Schwinger, E., Gal, A. Characterization of breakpoint sequences of five rearrangements in *L1CAM* and *ABCD1* (*ALD*) genes. *Hum Mutat* 19: pp. 526–35, 2002.

[134] Aubourg, P., Mosser, J., Douar, A.M., Sarde, C.O., Lopez, J., Mandel, J.L. Adrenoleukodystrophy gene: unexpected homology to a protein involved in peroxisome biogenesis. *Biochimie* 75: pp. 293–302, 1993.

[135] Poulos, A., Gibson, R., Sharp, P., Beckman, K., Grattan-Smith, P. Very long chain fatty acids in X-linked adrenoleukodystrophy brain after treatment with Lorenzo's oil. *Ann Neurol* 36: pp. 741–6, 1994.

[136] Leitner, H.M., Kachadourian, R., and Day, B.J. Harnessing Drug Resistance: Using ABC Transporter Proteins To Target Cancer Cells. *Biochem Pharmacol* 74(12): pp. 1677–85, 2007.

TITLES OF RELATED INTEREST

Colloquium Series on Stem Cell Biology

Titles Forthcoming in 2012

The Use of Induced Pluripotent Stem Cells to Study Genetic Diseases in Neuroscience
Ron Hart, Jennifer Moore, Mike Sheldon
Rutgers University

Neuronal Fate Specification of Pluripotent Stem Cells
Meng Li
MRC Clinical Sciences Centre, Imperial College London (UK)

Stem Cells and Progenitors in the Developing Liver
Marcus O. Muench
Blood Systems Research Institute, University of California, San Francisco

Signaling Mechanisms Controlling CNS Migration of Mesenchymal Stem Cells
Lei Zhang and Min Zhao
University of California, Davis

Stem Cell-based Therapy For Myelin Disorders
Sangita Biswas
Shriners Institute of Pediatric Regenerative Medicine

Systems Biology of Neural Stem Cells: Lessons from the Olfactory Epithelium
Anne L. Calof
University of California, Irvine

Homing of Stem Cells to Ischemic Myocardium

Jon C. George

Temple University School of Medicine

Emerging Roles for Neural Stem Cells in Memory, with an Emphasis on Emotional Memory

Daniela Kaufer

University of California, Berkeley

Neural Stem Cells in Hypoxic-Ischemic and Hemorrhagic Brain Injury in Newborns

Chia-Yi Kuan

Cincinnati Children's Hospital Medical Center

Stem Cell-based Therapy for Spinal Cord Injury

Ying Liu

Scripps Institute

The Biology and Therapeutic Potential of Stem Cells of the Oral Cavity

Sandu Pitaru

Tel Aviv University

For a full list of published and forthcoming titles:

http://www.morganclaypool.com/page/scb

Colloquium Series on Integrated Systems Physiology: From Molecule to Function to Disease

Editors

D. Neil Granger, Ph.D., *Boyd Professor and Head of the Department of Molecular and Cellular Physiology at the LSU Health Sciences Center, Shreveport*

Joey P. Granger, Ph.D., *Billy S. Guyton Distinguished Professor, Professor of Physiology and Medicine, Director of the Center for Excellence in Cardiovascular-Renal Research, and Dean of the School of Graduate Studies in the Health Sciences at the University of Mississippi Medical Center*

Physiology is a scientific discipline devoted to understanding the functions of the body. It addresses function at multiple levels, including molecular, cellular, organ, and system. An appreciation of the processes that occur at each level is necessary to understand function in health and the dysfunction associated with disease. Homeostasis and integration are fundamental principles of physiology that account for the relative constancy of organ processes and bodily function even in the face of substantial environmental changes. This constancy results from integrative, cooperative interactions of chemical and electrical signaling processes within and between cells, organs and systems. This eBook series on the broad field of physiology covers the major organ systems from an integrative perspective that addresses the molecular and cellular processes that contribute to homeostasis. Material on pathophysiology is also included throughout the eBooks. The state-of the art treatises were produced by leading experts in the field of physiology. Each eBook includes stand-alone information and is intended to be of value to students, scientists, and clinicians in the biomedical sciences. Since physiological concepts are an ever-changing work-in-progress, each contributor will have the opportunity to make periodic updates of the covered material.

For a full list of published and forthcoming titles:
http://www.morganclaypool.com/toc/isp/1/1

Colloquium Series on Developmental Biology

Editor

Daniel S. Kessler, Ph.D., *Associate Professor of Cell and Developmental Biology, Chair, Developmental, Stem Cell and Regenerative Biology Program of CAMB, University of Pennsylvania School of Medicine*

Developmental biology is in a period of extraordinary discovery and research. This field will have a broad impact on the biomedical sciences in the coming decades. Developmental Biology is interdisciplinary and involves the application of techniques and concepts from genetics, molecular biology, biochemistry, cell biology, and embryology to attack and understand complex developmental mechanisms in plants and animals, from fertilization to aging. Many of the same genes that regulate developmental processes underlie human regulatory gene disorders such as cancer and serve as the genetic basis of common human birth defects. An understanding of fundamental mechanisms of development is providing a basis for the design of gene and cellular therapies for the treatment of many human diseases. Of particular interest is the identification and study of stem cell populations, both natural and induced, which is opening new avenues of research in development, disease, and regenerative medicine. This eBook series is dedicated to providing mechanistic and conceptual insight into the broad field of Developmental Biology. Each eBook is intended to be of value to students, scientists and clinicians in the biomedical sciences.

For a full list of published and forthcoming titles:
http://www.morganclaypool.com/toc/deb/1/1

Colloquium Series on
The Developing Brain

Editor

Margaret McCarthy, PhD., *Professor, Department of Physiology; Associate Dean for Graduate Studies; and, Acting Chair, Department of Pharmacology & Experimental Therapeutics, University of Maryland School of Medicine*

The goal of this series is to provide a comprehensive state-of-the art overview of how the brain develops and those processes that affect it. Topics range from the fundamentals of axonal guidance and synaptogenesis prenatally to the influence of hormones, sex, stress, maternal care and injury during the early postnatal period to an additional critical period at puberty. Easily accessible expert reviews combine analyses of detailed cellular mechanisms with interpretations of significance and broader impact of the topic area on the field of neuroscience and the understanding of brain and behavior.

For a list of published and forthcoming titles:
http://www.morganclaypool.com/toc/dbr/1/1

Colloquium Series on Neuropeptides

Editors

Lakshmi Devi, Ph.D., *Professor, Department of Pharmacology and Systems Therapeutics, Associate Dean for Academic Enhancement and Mentoring, Mount Sinai School of Medicine, New York*

Lloyd D. Fricker, Ph.D., *Professor, Department of Molecular Pharmacology, Department of Neuroscience, Albert Einstein College of Medicine, New York*

Communication between cells is essential in all multicellular organisms, and even in many unicellular organisms. A variety of molecules are used for cell-cell signaling, including small molecules, proteins, and peptides. The term 'neuropeptide' refers specifically to peptides that function as neurotransmitters, and includes some peptides that also function in the endocrine system as peptide hormones. Neuropeptides represent the largest group of neurotransmitters, with hundreds of biologically active peptides and dozens of neuropeptide receptors known in mammalian systems, and many more peptides and receptors identified in invertebrate systems. In addition, a large number of peptides have been identified but not yet characterized in terms of function. The known functions of neuropeptides include a variety of physiological and behavioral processes such as feeding and body weight regulation, reproduction, anxiety, depression, pain, reward pathways, social behavior, and memory. This series will present the various neuropeptide systems and other aspects of neuropeptides (such as peptide biosynthesis), with individual volumes contributed by experts in the field.

For a list of published and forthcoming titles:
http://www.morganclaypool.com/toc/npe/1/1

Colloquium Series on
The Cell Biology of Medicine

Editor

Joel Pardee, Ph.D. *President, Neural Essence; formerly Associate Professor and Dean of Graduate Research, Weill Cornell School of Medicine*

In order to learn we must be able to remember, and in the world of science and medicine we remember what we envision, not what we hear. It is with this essential precept in mind that we offer the Cell Biology of Medicine series. Each book is written by faculty accomplished in teaching the scientific basis of disease to both graduate and medical students. In this modern age it has become abundantly clear that everyone is vastly interested in how our bodies work and what has gone wrong in disease. It is likewise evident that the only way to understand medicine is to engrave in our mind's eye a clear vision of the biological processes that give us the gift of life. In these lectures, we are dedicated to holding up for the viewer an insight into the biology behind the body. Each lecture demonstrates cell, tissue and organ function in health and disease. And it does so in a visually striking style. Left to its own devices, the mind will quite naturally remember the pictures. Enjoy the show.

For a list of published and forthcoming titles:
http://www.morganclaypool.com/toc/cbm/1/1

Colloquium Series on
Genomic and Molecular Medicine

Editor

Professor Dhavendra Kumar, MD, FRCP, FRCPCH, FACMG, *Consultant in Clinical Genetics, All Wales Medical Genetics Service Genomic Policy Unit, The University of Glamorgan, UK Institute of Medical Genetics, Cardiff University School of Medicine, University Hospital of Wales*

The progress of medicine has always been driven by advances in science and technology. The practice of medicine at a given place and time is a reflection of the current knowledge, applications of the available information and evidence, social/ cultural/ religious beliefs and statutory requirements. Thus it needs to be dynamic and flexible to accommodate changes and new developments in basic and applied science in keeping with the individual and societal expectations.

From 1970 onwards, there has been a continuous and growing recognition of the molecular basis of medical practice. Most medical curricula allow sufficient space and time to ensure satisfactory coverage of the basic principles of molecular biology. Emphasis is given on the relevance of molecular science in the practice of clinical medicine paving the way for a more holistic approach to patient care utilizing new dimensions in diagnostics and therapeutics. It is extremely important that both teachers and students have an agreed agenda for learning and applications of molecular medicine. This should not be restricted to few uncommon genetic conditions but extended to include inflammatory conditions, infectious diseases, cancer and age-related degenerative conditions involving multiple body systems.

Alongside the developments and progress in molecular medicine, rapid and new discoveries in genetics led to an entirely new approach to the practice of clinical medicine. However, the field of genetic medicine has been restricted to the diagnosis, offering explanation and assistance to patients and clinicians in dealing with a number of relatively uncommon inherited disorders. Nevertheless this field has gradually established and accorded the specialty status in the medical curriculum of several countries.

Since the completion of the human genome in 2003 and several other genomes, there is now plethora of information available that has attracted the attention of molecular biologists and allied researchers. A new biological science of Genomics is now with us with far reaching dimensions and applications. Rapid and revealing findings in genomics have led to changes in the fundamental concepts in cell and molecular biology. A present day student of biology is expected to conceptualize the sequence of genome-gene-molecule-cell with reference to specific tissue, organ and a body system. In other words evolutionary and morbid changes at the genome level could be the basis of normal human variation and disease. During the last decade, rapid progress has been made in new genome-level diagnostic and prognostic laboratory methods. Applications of individual genomic information in clinical medicine have led to the prospect of robust evidence-based personalized medicine. Genomics has led to the discovery and development of a number of new drugs with far reaching implications in pharmaco-therapeutics. The existence of Genomic Medicine around us that is inseparable from molecular medicine. Both genomic and molecular medicines are in fact two dimensions of the integrated modern molecular medicine with tremendous implications for the future of clinical medicine.

For a full list of published and forthcoming titles:
http://www.morganclaypool.com/page/gmm

Colloquium Series on
The Genetic Basis of Human Disease

Editor

Michael Dean, Ph.D., *Head, Human Genetics Section, Senior Investigator, Laboratory of Experimental Immunology National Cancer Institute (at Frederick)*

This series will explore the genetic basis of human disease, documenting the molecular basis for rare, common, Mendelian, and complex conditions. The series will overview the fundamental principles in understanding such as Mendel's laws of inheritance, and genetic mapping through modern examples. In addition current methods (GWAS, genome sequencing) and hot topics (epigenetics, imprinting) will be introduced through examples of specific diseases.

For a full list of published and forthcoming titles:
http://www.morganclaypool.com/page/gbhd

Colloquium Series on Systems Biology and Data Integration

Editors

Aristotelis Tsirigos, Ph.D., *Research Scientist, IBM Computational Biology Center, IBM Research*
Gustavo Stolovitzky, Ph.D., *Manager, Functional Genomics and Systems Biology, IBM Computational Biology Center, IBM Research*

Since the beginning of the 21st century, the development of high-throughput techniques has accelerated discovery in biology. The influx of an unprecedented amount of data presents challenges as well as great opportunities for improving our understanding of living systems. Systems Biology is an interdisciplinary field which, by integrating diverse types of data, such as Genomics, Epigenomics, Proteomics, Metabolomics, etc., aims at modeling biological processes at a systems level - tissue, organ, organism - both in normal function and under stress. This eBook series is dedicated to an in-depth presentation of topics in Systems Biology of research concerning human disease. Each book is written by an expert research scientist who has extensive experience in a particular model system of disease and has demonstrated in his work the value of integrating multiple types of data with potential practical therapeutic applications. Its intended audience is students and scientists in the biomedical sciences who are interested in participating in this fascinating research game of discovery.

For a full list of published and forthcoming titles:
http://www.morganclaypool.com/page/sbdi

Colloquium Series on Molecular Mechanisms in Critical Care Medicine

Editor

Lew Romer, M.D., *Associate Professor of Anesthesiology and Critical Care Medicine, Cell Biology, Biomedical Engineering, and Pediatrics, Center for Cell Dynamics, Johns Hopkins University School of Medicine*

The idea is to provide researchers and critical care fellows with the context and tools to address the burgeoning data stream on the genetic and molecular basis of critical illness. This will be clustered around classical systems including CNS, Respiratory, Cardiovascular, Hepatic, Renal, and Hematologic.

We will explore major themes and categories of current research and practice, such as: to identify genes and molecules involved in control mechanisms for homeostasis and injury response; to identify developmental processes in which these genetic and molecular control mechanisms have a role; and to identify diseases in which these genetic and molecular control mechanisms go awry. The Series will also seek to explain and anticipate mechanisms by which critical care interventions modulate or exacerbate dysfunction of these genes and molecules. Similarly, this Series will seek to describe the role of genotypic variation in explaining the variable phenotypes of critical illness and the interaction between genetic and environmental factors in producing the variable phenotype of critical illness.

For a full list of published and forthcoming titles:
http://www.morganclaypool.com/page/mmccm

Colloquium Series on Protein Activation and Cancer

Editor

Majid Khatib, Ph.D., *Research Director, INSERM, and, University of Bordeaux*

This series is designed to summarize all aspect of protein maturation by the convertases in cancer. Included topics are dealing with the importance of these processes in the acquisition of the malignant phenotype by tumor cells, induction of tumor growth and metastasis. This series also provide latest knowledge on the clinical significance of convertases expression and activity and their protein substrates maturation in various cancers. The potential use of their inhibition as a therapeutic approach is also explored.

Titles Forthcoming in 2012

Animal Models of Proprotein Convertases-mediated Carcinogenesis
Andres Klein-Szanto and Daniel E. Bassi
Fox Chase Cancer Center

The Role of Proprotein Convertases During Development of Human Gynecological Cancers
Daniel E. Bassi and Andres Klein-Szanto
Fox Chase Cancer Center

The Role of Proprotein Convertases in Cancer Progression and Metastasis
Peter Metrakos
McGill University

Modulators of Proprotein Convertase Subtilisin Kexins: Design, Cellular Delivery and Therapeutic Implications
Ajoy Basak
University of Ottawa

Genetic Diseases Involving Neuropeptide Systems
John Creemers
Katholieke Universiteit Leuven, Belgium

For a full list of published and forthcoming titles:
http://www.morganclaypool.com/page/pac

Colloquium Series on
The Building Blocks of the Cell

Editor

Ivan Robert Nabi, Ph.D. *Professor, University of British Columbia, Department of Cellular and Physiological Sciences*

This Series is a comprehensive, in-depth review of the key elements of cell biology including 14 different categories, such as Organelles, Signaling, and Adhesion. All important elements and interactions of the cell will be covered, giving the reader a comprehensive, accessible, authoritative overview of cell biology. All authors are internationally renowned experts in their area. The Series will include over 50 ebooks.

Titles Forthcoming in 2012

Actin Cytoskeleton
Jonathan M. Lee
University of Ottawa

Autophagy
Vojo Deretic
University of New Mexico Health Science Center

Calcium Signaling
James D. Johnson
University of British Columbia

Caveolae
Radu V. Stan
Dartmouth Medical School

Cellular Adhesions

Jacky G. Goetz

Institute of Genetics and Molecular and Cellular Biology (IGBMC), France

Cilia

Lynne Quarmby

Simon Fraser University

Endocytosis

Christophe Lamaze

INSERM

Epithelial Polarity

Gerard L. Apodaca

University of Pittsburgh School of Medicine

Extracellular Matrix

Hakima Moukhle

University of British Columbia

Gene Transcription

T. Michael Underhill

University of British Columbia

Glycosylation

James W. Dennis

Samuel Lunenfeld Research Institute, University of Toronto

Imaging the Cell

Edwin Moore and David Scriven

University of British Columbia

Intermediate Filaments

Normand Marceau

Centre de Recherche du CHUQ (Pav. L'Hôtel-Dieu de Québec)

Intracellular Lipid Traffic

Christopher J.R. Loewen

University of British Columbia

Ion Transport
Christopher Ahern
University of British Columbia

Lipid Rafts
John R. Silvius
McGill University

Mitochondria
Andreas Reichert, Goethe University Frankfurt Medical School
Michael Zick, Dartmouth Medical School

Phagocytosis
Rene E. Harrison
University of Toronto Scarborough

Planar Cell Polarity
Calvin D. Roskelley
University of British Columbia

Plasma Membrane Domains
Akihiro Kusumi, Takahiro Fujiwara and Ziya Kalay, Rinshi Kasai
Institute for Integrated Cell-Material Sciences (iCeMS), Kyoto University

Unfolded Protein Response and Endoplasmic Reticulum-Associated Degradation (ERAD)
Erik Lee Snapp
Albert Einstein College of Medicine of Yeshiva University

For a full list of published and forthcoming titles:
http://www.morganclaypool.com/page/bbc